PUNU

Yankunytjatjara plant use

Traditional methods of preparing foods, medicines, utensils and weapons from native plants

Contributors

Pompey Everard, Mollie Everard, Murika, Tommy Tjampu, Kanytji, Sam Pumani, Milatjari

Compiled and edited by

Cliff Goddard, Arpad Kalotas

jukurrpa books

First published in 1985
New edition published in 2002 by
jukurrpa books,
an imprint of

IAD Press
PO Box 2531
Alice Springs NT 0871
Ph: (08) 8951 1334
Fax: (08) 8952 2527
iadpress@ozemail.com.au

National Library of Australia Cataloguing-in-Publication entry:

Punu: Yankunytjatjara plant methods of preparing foods, medicines, utensils and weapons from native plants.

New ed.
Includes bibliography
ISBN 1 86465 036 2.

1. Plants, Useful — Australia. 2. Yankunytjatjara (Australian people) — Medicine. 3. Yankunytjatjara (Australian people) — Ethnobotany. 4. Yankunytjatjara (Australian people) — Food. I. Kalotas, Arpad C. II. Goddard, Cliff. III. Everard, Pompey.

581.630994

Illustrations by Julie Jones, with additional illustrations by Jill Hodson, Mary Flynn
Heather Carey and Christine Bruderlin
Design and layout by Louise Wellington
Printed in Australia by Hyde Park Press

ATSIC

IAD Press acknowledges the assistance of the Federal Government's Language Access Initiatives Program (administered by ATSIC) in the publication of this book.

ACKNOWLEDGEMENTS

Our thanks go to the staff of the Northern Territory Herbarium, Alice Springs (Botany Unit, Conservation Commission of the Northern Territory) for providing access to their botanical collections for illustration purposes.

Thanks to Jenny Curnow for the initial typing of the text.

TRANSLATOR'S NOTE

Probably there is no such thing as the perfect translation, certainly not in this book anyway. A good translation is one that is good for the purpose or person it is intended for.

The translations in this book try to do two things. To convey the content and the tone of the contributors' texts as well as possible to someone with little or no knowledge of Yankunytjatjara, and to be as useful as possible for learners of the language as well. These purposes sometimes conflict, and there are additional problems that come from the fact that in written form the stories lack the performance quality that was so much a part of the way they were told.

Each translation has passed through several revisions, and most still need improvement. In doing the translations, my main consultant has been Mr Yami Lester, who deserves to be seen as a major contributor to this book, along with his relatives at Mimili.

Cliff Goddard

Contents

Translator's Note ... iv

Preface ... vi

Introduction ... 1

The region ... 1

Climate and landforms ... 2

Vegetation ... 3

Plants and people ... 6

Vegetable food: ***mai*** ... 8

Fruits ... 8

Seeds ... 9

Roots, greens, galls and fungi ... 11

Nectars, sweet secretions and gums ... 12

Medicinal plants: ***pu<u>n</u>u ngangka<u>r</u>i*** ... 13

Other plant uses ... 14

The language of the stories ... 16

Vocabulary ... 17

Grammar ... 19

Sound system ... 21

Yankunytjatjara Plant Use ... 25

Trees ... 25

Shrubs ... 40

Subshrubs ... 61

Herbs ... 65

Vines, fungi and mistletoes ... 72

Grasses ... 81

Glossaries ... 89

Plant parts and related terms ... 89

Habitats and vegetation ... 91

Processing plants (stages and products, processes and implements) ... 92

Lists of Plants ... 96

Annotated list of additional plants ... 96

Alphabetical list under botanical name ... 105

References and Resources ... 111

Index ... 114

PREFACE

For generations the Yankunytjatjara people relied for survival on a deeply practical knowledge of their environment. A major part of this knowledge concerned plants, which provided them with, among other things, food, medicine, and the raw material for utensils and weapons. This book, based on contributions from the people of Mimili, aims to provide a resource on the traditional knowledge and use of plants by the Yankunytjatjara people.

The heart of the book is a set of detailed treatments of individual plants, chosen so as to span both the range of plant forms that occur in the Yankunytjatjara region, and the diverse uses to which they are put. The old people of Mimili have provided one or more texts, or stories, for each of these plants, describing its preparation and use. These stories are presented with English translations, an English summary of the traditional knowledge, and a description of the plant and its habitat. Also included are a set of glossaries, listing many of the specialised terms Yankunytjatjara people use when describing plants and plant use, an annotated checklist of some eighty or so other species, and a general introduction.

The Aboriginal contributors to this book are justly proud of their traditional knowledge, and concerned that their children retain as much of it as possible, given that economic relationships with the land have been drastically simplified in recent times. We hope the stories in this book will provide reading material in schools, and that teachers will incorporate the information into the curriculum in various ways.

The old people are also keenly aware of the contrast in community health between present and traditional lifestyles. Hopefully health workers will find the information about traditional diet and medicine of some use. We also hope that language learners and other people interested in Aboriginal culture will find the book helpful and interesting.

It goes without saying that this work is not exhaustive. We should also point out that it will inevitably contain errors and inelegancies, especially in the translations and glossaries.

We have gained much satisfaction in putting this book together. We hope the contributors and you, the reader, will enjoy it too.

Arpad Kalotas — botanist
Cliff Goddard — linguist

INTRODUCTION

The region

The traditional country of the Yankunytjatjara extends from Uluṟu (Ayers Rock) in the north, south-east through the Musgrave and Everard Ranges to the Great Victoria Desert (see map below). Since 1981 the central part of this area has been Aboriginal freehold land, under the South Australian Pitjantjatjara Land Rights Act.

The environmental characteristics described here are specifically for the Everard Ranges, where the Yankunytjatjara contributors to this book live. However, they may be applied generally to the whole region.

The main population centre in the Everards is Mimili, previously known as Everard Park, a small community of mainly Yankunytjatjara-speaking people situated in the south of the ranges. It is approximately 1000 km north-west of Adelaide, and about 70 km west of Indulkana (Iwantja), itself a few kilometres west of the Stuart Highway and the railway line.

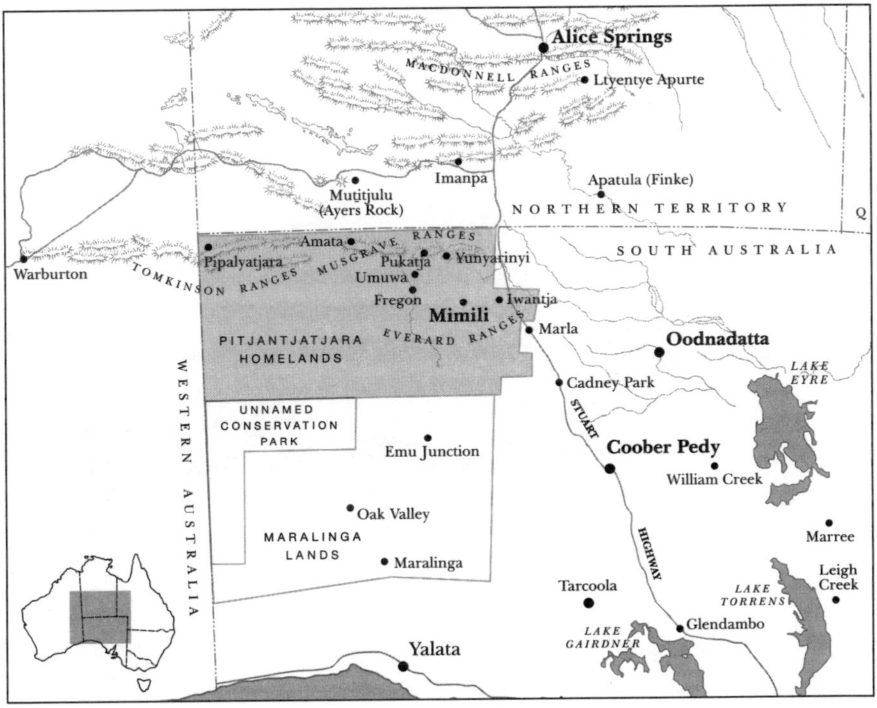

Climate and landforms

The Everard Ranges lie in one of the most arid regions of Australia. The average rainfall is less than 150 mm. Rains may occur at any time of the year, due to the competing influences of the temperate weather patterns of the south of Australia, and the sub-tropical and monsoonal patterns to the north.

Daytime temperatures are consistently very hot in summer (a mean daily maximum of 37°C in January) and pleasantly temperate in winter (a mean daily maximum of 20°C in July). In all seasons the sky is generally cloudless, with a daily average of over nine hours of bright sunshine. Nights are temperate in summer and cold in winter, with frequent frosts.

The country around the Everards varies from rugged rangeland, often with magnificent rounded granite outcrops (known as domed inselbergs), to the sandplains and dunefields of the open country. Limestone rises and salt lakes are less common landforms. There are no permanently flowing rivers in the region, though watercourses exist and drain water after rain. In the hills and ranges, and among rocky outcrops, these take the form of gullies. Creeks and relatively large rivers dissect the plains and dune regions.

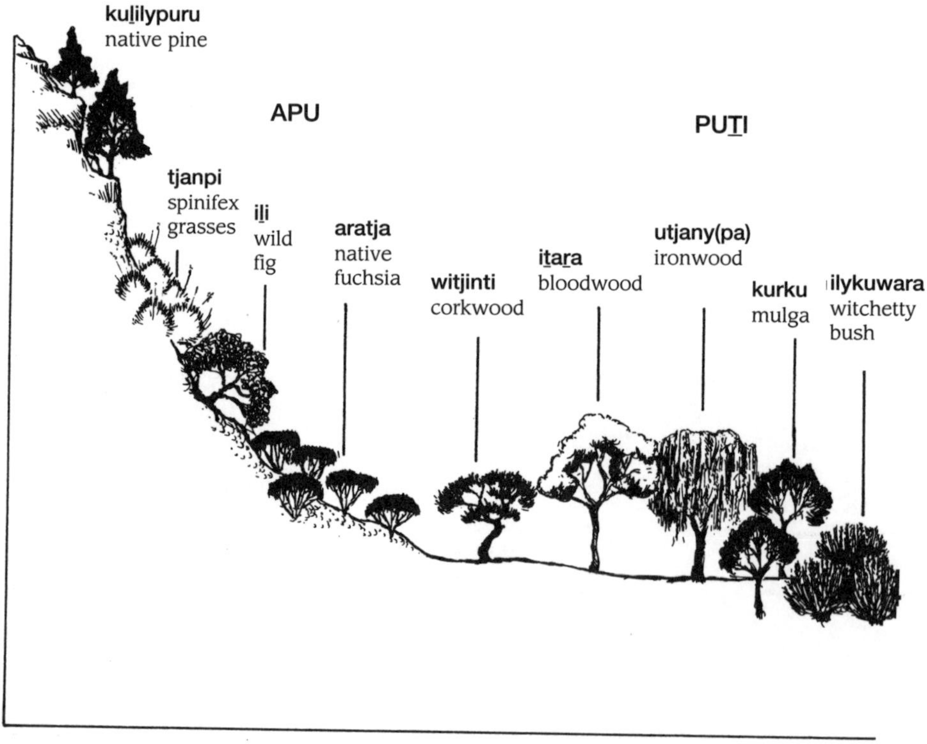

Vegetation

The vegetation consists of a variety of plant forms (including trees, shrubs, grasses, herbs and vines) which make up communities of a relatively simple structure, typical of much of Central Australia. Common formations include woodlands, shrublands, grasslands with scattered trees and shrubs, and largely uninterrupted grasslands. The nature of the vegetation is often related to the habitats or landforms they occupy. The vegetation around the Everards is described below in terms of the major habitats recognised by Yankunytjatjara people.

Apu rocky outcrops, hills, ranges

Apu are sparsely vegetated by spinifex ***tjanpi*** and various acacias or wattles. In the Everards (but not generally elsewhere in the region) the dominant species over rocky hills is *Acacia olgana* ***kaliwara***. Other species characteristically found on ***apu*** are the wild fig ***ili*** and medicinal plants such as the native fuchsias ***aratja*** and ***irmangka-irmangka***, caustic vine ***ipi-ipi***, mint bush ***karingana*** and hop bush ***tjininy(pa)***.

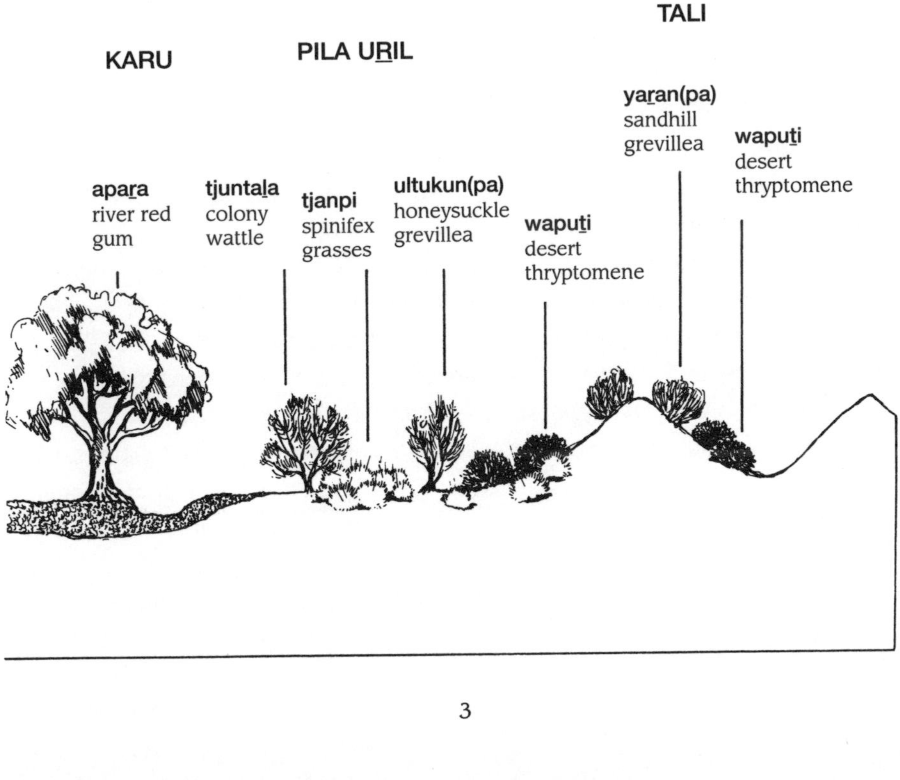

Spear bush ***urtjan(pa)***, native pine ***kulilypuru***, the herbaceous wild tobacco ***pulyantu*** and ***mingkul(pa)*** and rock isotome ***tjuntiwari*** are also found almost exclusively on, or around, ***apu***.

Karu watercourses (gullies, creeks and rivers)

The major watercourses of the region are wide, sandy riverbeds, often with adjacent floodplains. They are dry for most of the year, though after heavy summer rains they may become raging rivers. River red gum ***apara*** is a characteristic and dominant element fringing watercourses, or occurring throughout the riverbed, forming tall open woodlands.

A dominant species in rocky gullies in particular is the inland ti-tree ***ilpili***, which forms tall shrublands. Other species from surrounding habitats may be found in watercourses, particularly acacias (wattle) such as dead finish ***kurara*** and umbrella bush ***watarka***. The hop bush ***tjininy(pa)*** and the wild plum ***kupata*** are also occasionally found in ***karu***.

Herbs and grasses thrive in the groundstorey in good seasons, and include such plants as the bush onion ***tjanmata*** and blue rod ***intiyanu***. Sedges and tall grasses, collectively referred to as ***puta-puta*** and ***ilintji*** respectively, make up the groundstorey.

Puti woodlands, shrublands

Pila spinifex grasslands, plains

Uril open country, grasslands, plains

A variety of vegetation formations are found on sandplains, which are the most extensive habitats in the region. The variation is largely determined by the change in soil types between areas adjacent to ranges, as compared to those further away. ***Puti***, ***pila*** and ***uril*** refer to the most important of the sandplain vegetation formations.

Puti are the woodlands and shrublands. Most common, and dominant in some areas, are woodlands of mulga ***kurku***. They may be closed, open or sparse formations, with a ground layer of scattered perennial grasses, herbs and subshrubs. In certain areas, spinifex is the dominant groundcover.

Common in open and sparse mulga formations are the trees corkwood ***witjinti***, ironwood ***utjany(pa)***, bloodwood ***itara*** and silver-leaved water bush ***ilpara***. Shrubs such as the emu bush ***tulypur(pa)***, the fuchsia bush ***ngarankura***, ***mintjingka***, various cassias (sennas) ***punti***, and the acacias prickly wattle ***ngatun(pa)***, dead finish ***kurara***, umbrella bush ***watarka*** and the witchetty bush ***ilykuwara*** are also often associated with mulga. Other trees and shrubs occasionally found here include the slender bush currant

pakali-pakali and the bush bean vine ***wintjulany(pa)***, both food species. In some areas the trees and shrubs associated with mulga may form communities virtually independent of mulga. For example, ***witjinti***, ***utjany(pa)*** and ***itara*** may each form open woodlands, though more often they occur together in open mixed woodlands.

Over areas of flat or slightly undulating sandplains, spinifex forms hummock grassland ***pila***. Mallee (multi-stemmed eucalypts), commonly called ***altar(pa)*** in Yankunytjatjara, may dominate spinifex sandplains, and with other trees and shrubs form mixed woodlands. Some plants, such as the honeysuckle grevillea ***ultukun(pa)***, may form pure stands over spinifex, and may also be found scattered with other trees and shrubs. Other species commonly found on spinifex plains include the shrubs wild plum ***kupata***, witchetty bush ***ilykuwara***, colony wattle ***tjuntala***, umbrella bush ***watarka***, hop bush ***tjininy(pa)***, cassias (sennas) ***punti*** and desert poplar ***kaluti***.

The groundstorey of the sandplains is, in good seasons, a mixture of woody subshrubs, grasses and herbs. Important food plants common after rains include the 'bush tomatoes' desert raisin ***kampurara*** and western nightshade ***ituny(pa)***; the grasses naked woollybutt ***wangunu*** and native millet ***kaltu-kaltu***; herbs such as rats' tails ***kalpari***, the creeper ***anultja*** and the succulents parakeelya ***parkily(pa)*** and inland pigweed ***wakati***; and even edible fungi such as the native truffle ***witita***.

Tali sandhills, sand-dunes, sandhill country

Tali are a dominant habitat throughout the region, particularly to the south, in the Great Victoria Desert. They support a characteristic vegetation of open or sparse, tall or low shrublands with spinifex as the dominant groundcover. Common plant species include sandhill grevillea ***yaran(pa)***, desert thryptomene ***waputi***, wheel fruit ***untalya*** and a native fuchsia ***tjutu***. Other species common in areas between sandhills, and on the open plains, such as the honeysuckle grevillea ***ultukun(pa)*** and the emu poison bush ***tjila***, ***walkal(pa)***, are also found on sandhills.

Plants and people

The detailed knowledge the Yankunytjatjara have of plants and their uses forms the basis for the subsistence activities of hunting and gathering. Plants provide the various vegetable foods gathered, and act as hosts for insects which produce sweet secretions, edible galls and edible larvae ***maku***, 'witchetty grubs'. Plants also provide the raw materials for hunting weapons, the implements used in gathering, and many other items of material culture.

Though the living situation at Mimili is no longer fully 'traditional', hunting and gathering remain important economic activities. People engage in them to supplement store-purchased foods, for teaching purposes, and just pure enjoyment. Not all plant foods discussed below are gathered today. Generally the readily accessible foods, such as fruits, some roots, honeys and galls, are still eagerly foraged for. The manufacture of tools from wood is still practised, not so much for the purposes of hunting and gathering, but to generate income from their sale as artefacts. Native tobacco is highly prized as a chewing tobacco. The use of plants for firewood and shelter is essential in everyday life.

The ethnobotanical information presented in this book is derived largely from the experiences of Yankunytjatara people living at Mimili. This is highlighted in their stories, which feature in Part 2. References dealing specifically with Yankunytjatjara plant use from other places have also been drawn upon. It should be noted, however, that this book is far from complete: there are several major plant species known to the Yankunytjatjara which are not covered at all. Furthermore, the focus on Mimili alone precludes inclusion of species whose distribution limits lie outside the Everard Ranges area, for instance the desert kurrajong ***ngalta*** and the desert oak ***kurkara***. (Some information on these species is available in the literature on Pitjantjatjara plant use, listed in Part 5.)

The Yankunytjatjara classify food into two main categories: ***kuka*** meat, and ***mai*** vegetable food. The word ***mai*** is commonly used in front of the name of particular food species, for example the wild fig ***ili*** is often referred to as ***mai ili***. Additional categories are ***maku*** edible larvae, ***wama***, ***tjuratja*** nectars and other sweet substances, and ***tjau*** edible gums produced by some plants.

In Yankunytjatjara all plants may be called ***punu***, though the core meaning of this word refers to larger plant types, such as trees and shrubs. (***Punu*** can also be used to refer to sticks and to things made of wood, such as bowls, spears and so on.) ***Ukiri*** refers to any green plant, though it is usually applied to green grasses and herbs, including succulents and vines. Grasses are called ***putja***.

Economically important plants almost always have at least one specific name, sometimes more. For instance, the native millet is known by two specific names ***kal̲tu-kal̲tu*** and ***kutja***. The river red gum is known by the specific name ***apar̲a***, though it can also be referred to as ***it̲ar̲a***, a general term for large eucalypts, which applies also to the bloodwood. One of the most widespread and economically important species is mulga (*Acacia aneura*), which has an unusually wide variation in form. It may be shrubby or tree-like; leaves may be needle-like or flattened, straight or curved. The Yankunytjatjara recognise the relatedness of these forms by referring to them all as ***kurku***, but they distinguish as many as five forms by means of additional specific names.

Many less important plant species have no specific name. For example, ***put̲a-put̲a*** is used for at least four species of sedge, and ***ilintji*** for at least four species of tall grass. General terms like these group together plants which are very similar to one another, in a similar fashion to English terms such as sedge, wattle, gum tree and so on. Some of the general terms such as ***witjinti*** (corkwoods), and ***altar(pa)*** (mallees) apply to species of the same botanical genus, Hakea and Eucalyptus respectively for these examples. Other general terms may apply to species of the same family, rather than genus. Examples include ***put̲a-put̲a*** for members of the sedge family Cyperaceae and ***ilintji*** for tall grasses in the Poaceae. Other general terms do not correspond to the botanical system of classification, since they include species from different families. For example, ***tjul̲pun-tjul̲pun(pa)*** is a word very similar to the English 'wildflower', and so takes in species of daisy and pussytail, from different botanical families.

In most cases, plant names are not descriptive, and their origins are unknown. Some interesting exceptions are ***ipi-ipi*** caustic vine, ***tjilka-tjilka*** and ***mukul-mukul(pa)***: ***ipi*** means 'breast milk', and the caustic vine has a milky sap; ***tjilka*** means 'prickle, thorn', and the two species of ***tjilka-tjilka*** are both prickly; ***mukul(pa)*** means 'hook, barb', and the leaves of this plant are hook-shaped.

In the reduplicated nature of their names, these examples illustrate another common feature of Yankunytjatjara plant names. However, most reduplicated plant names differ from these in that the reduplicated part is not generally a word in its own right, or if it is, there is no apparent relation between this word and the plant name. For instance, ***tjul̲pun-tjul̲pun(pa)*** means 'wildflower', but there is no word ***tjul̲pun(pa)***; ***pakal̲i-pakal̲i*** means 'bush currant', but the word ***pakal̲i*** 'grandson' bears no apparent relation to the plant.

Vegetable food: mai

Mai refers to the various vegetable plant foods such as fruits, seeds, leafy greens, roots, insect galls and fungi.

Fruits

Nowadays the most used plant foods are the fruits, because they require the least preparation. Several fruits, particularly the quandong ***mangaṯa*** and wild fig ***iḻi***, are distributed widely enough, and occur in sufficient quantities when in season, to make them major food sources ***mai puḻka***. The dried fruits of these two plants can be reconstituted with water, an important feature when other food was scarce. The wild plum ***kupaṯa*** is also fairly widespread, and quite abundant when in season. Also common and much sought after is the desert raisin ***kampuṟara***, which tends to remain on the small bushes for a long time in the dried 'raisin' stage. They may be pounded,

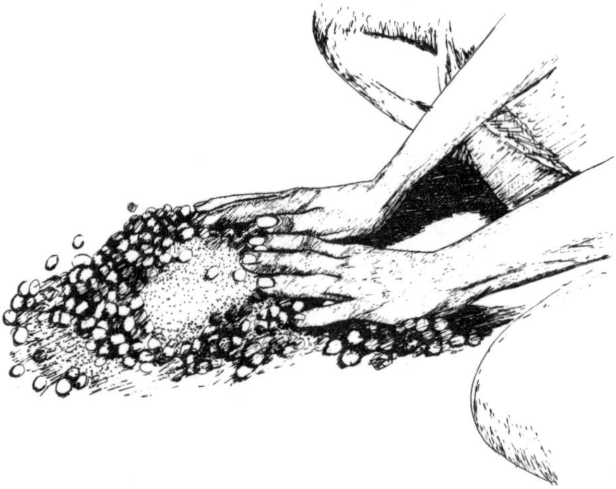

Kuṟapunganyi*: rubbing desert raisins* ***kampuṟara*** *against the ground to prevent headaches which otherwise result from eating too many of the berries.*

moistened and made into balls which can be stored and eaten at leisure.

Of the remaining bush fruits, the bush currant ***pakaḻi-pakaḻi*** is a slender, narrow-leaved shrub occurring only occasionally, in association with mulga ***kurku***. The wild gooseberry ***kulypur(pa)***, ***tawal-tawal(pa)*** is eaten when available. The western nightshade ***iṯuny(pa)***, another 'bush tomato', is not a favourite food because of its bitter juice. However, if the juice is squeezed out, it can be eaten raw or baked. The small sticky fruits of the mistletoes are also eaten; they are regarded as ***tjitjiku mai*** children's food.

These fruits all have some nutritional value and help make up a balanced diet. For instance, the quandong ***mangaṯa*** and bush tomatoes such as ***kampuṟara*** are rich in vitamin C; the wild fig ***iḻi*** has high levels of calcium (Peterson 1978).

FRUITS

Trees	*iḻi* *mangaṯa*	wild fig quandong *
Shrubs	*pakaḻi-pakaḻi* *kupaṯa*	bush currant wild plum *
Subshrubs	*kampuṟara* *kulypur(pa),* *tawal-tawal(pa)* *iṯuny(pa)*	desert raisin wild gooseberry western nightshade *
Vine	*wintjulany(pa)*	bush bean
Mistletoes	*ngaṉtja,* *parka-parka*	—

* treated in detail in Part 2

Seeds

Seed foods ***mai kaḻka, uḻiṉy(pa)*** were an important part of the Yankunytjatjara diet, largely due to their high protein values (Peterson 1978). Because of the lengthy and often arduous nature of the collecting and processing required, and the current availability of wheat flour, the preparation of seed food is no longer regularly engaged in, except for teaching and demonstration purposes.

The species traditionally utilised vary in habit or growth form: some are trees or tall shrubs, mainly wattles (acacia species), others are grasses or herbs. Generally the mature, hard seeds of trees and tall shrubs were gathered from the dried-out pods on the plants, or from the fallen pods in the leaf litter at the base of the plants. The seeds were then separated from the pods and other unwanted plant pieces by threshing, yandying (shaking in a dish) and winnowing. In many cases they were then parched in hot ashes and sand in a ***kaṉil(pa)*** dish. After this, they were cleaned again by yandying, sprinkled with water, and ground to a paste ***latja*** which was eaten without further preparation. There were some variations to the processes as outlined here, some of which are highlighted in some of the stories in Part 2.

Seeds of the grasses and the rats' tails herb ***kalpar̲i*** were collected by rubbing them by hand from the seedheads of the plant. Grass seeds required further attention to separate the seed from the outer covering (glumes), by rubbing, pounding, and in some cases singeing the seedheads. The cleaned seed was then ground with water into a paste or slurry which was baked in hot ashes and earth to produce a seedcake or traditional damper ***nyuma***, ***wanytji***.

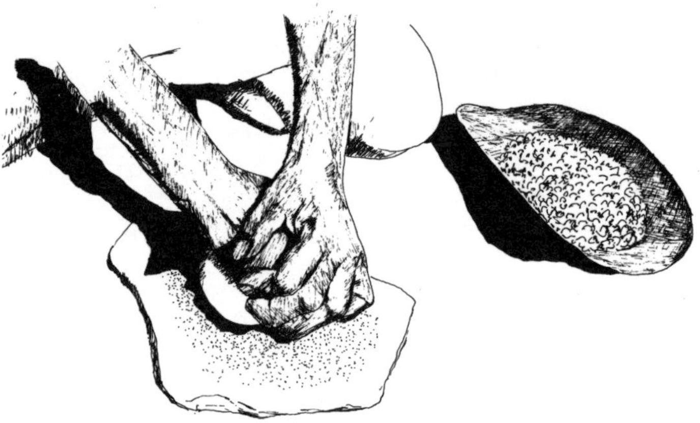

In all instances of seed processing, a hand-held millstone ***tjungar̲i*** and a flat, often grooved lower grindstone ***tjiwa*** were used in grinding the seed. These tools are the exclusive property of women (see Hamilton 1980 for more detail).

Seeds

Trees or shrubs	***kurku, wintalyka***	mulga
	kur̲ara	dead finish
	tjuntal̲a	colony wattle
	ngat̲un̲(pa)	prickly wattle
	kal̲iwar̲a	—
Grasses	***wangun̲u***	naked woollybutt
	kal̲tu-kal̲tu	native millet
Herbs	***wakati***	pigweed
	kalpar̲i	rats' tails

Roots, greens, galls and fungi

Root ***ungka*** foods vary from those foraged whenever available to those used only in case of emergency. The bush onion ***tjanamaṯa***, the most commonly mentioned root food, is a sedge with many small underground bulbs. It occurs throughout the region in and along watercourses, and nowadays is a popular snack food. It can be easily dug out by hand and, once the outer papery covering has been removed by rubbing the onions between the palms, may be eaten raw or lightly roasted.

The roots of the drought-resistant tarvine ***pilyaḻi*** were an emergency food, resorted to when other food was scarce. It is a flat, spreading herb, originating from a single taproot, much like a carrot, as people point out. It is found in a variety of habitats, especially along watercourses. It was dug from the ground and lightly roasted.

Vegetable greens ***ukiṟi*** that were traditionally used included native cress ***unmuṯa*** and the succulent parakeelya ***parkily(pa)***. Other vegetable foods include the insect galls ***tarulka*** mulga apple, which forms on mulga, and ***anguṟa***, which forms on the bloodwood tree, and the native truffle ***witiṯa***.

***anguṟa**: bloodwood gall*

ROOTS, GREENS, GALLS AND FUNGUS

Sedge	*tjanmaṯa*	bush onion
Herbs	*pilyaḻi*	tarvine
	unmuṯa	native cress
	parkily(pa)	parakeelya
Insect galls	*tarulka*	mulga apple
	anguṟa	bloodwood apple
Fungus	*witiṯa*	native truffle

Nectars, sweet secretions and gums

Honey plants, as they are commonly known in English, are often a ready source of nectar. Honeysuckle grevillea **ul̲tukun(pa)** and corkwood **witjinti** are the most readily accessible; the nectar can be sucked directly from the compound flowerheads or soaked in water to make a sweet drink. The flowers of **mintjingka** native fuchsia can be sucked individually. Gathering the nectar and dew of the desert thryptomone **waput̲i** requires some skill in beating the flowering bushes with a collecting dish.

Some plants bear sweet substances as a result of insect activity. The most well-known is **ngapar̲i**, a white, flaky crust found on the leaves of the river red gum at certain times of the year. It is a secretion of plant sugars produced by minute sap-sucking scale insects. It can be slipped off the leaves and eaten; to get larger quantities, whole branches are broken off, left to dry in the sun, then shaken, and the dried flakes are made into balls. The other commonly occurring secretion is **kurku** or **kurkunytjungu**, a mulga honeydew produced by a red lac scale insect. It is a clear, sticky liquid that appears as droplets on the stems of mulga. The branchlets are either sucked directly, or soaked in water to make a sweet drink.

Certain plants, such as the ironwood **utjany(pa)**, exude edible gum **tjau** from their trunks or branches. These are a kind of natural lolly, often sought out by children.

NECTARS, SWEET SECRETIONS AND GUMS

Trees	*witjinti*	corkwoods
	utjany(pa)	ironwood
	apar̲a	river red gum
	kurku	mulga
Tall shrubs	*ul̲tukun(pa)*	honeysuckle grevillea
	mintjingka, ngar̲ankur̲a	native fuchsia
Shrub	*waput̲i*	desert thryptomene

Medicinal plants: pu<u>n</u>u ngangka<u>r</u>i

A variety of medicinal plants occur throughout the region. Depending on locality, different plant species may be used to treat the same ailments. Most medicinals were used to make ointments, liniments or washes. In most cases the leafy parts were pounded and mashed with water and rubbed over the body, particularly the chest and head, for general illnesses. In the case of the quandong ***manga<u>t</u>a*** the kernel of the nut of the fruit is used to make an oily paste which is rubbed into the body for aches and pains. In addition, poultices of chopped up herbs were often applied to the head, enclosed in grass and tied in place.

Some plants, such as the native fuchsia ***aratja***, caustic vine ***ipi-ipi***, the turpentine bush ***munyu<u>n</u>(pa)*** and the hop bush ***tjininy(pa)***, were used in a process of fumigation, or smoke treatment. It is likely this was reserved for serious complaints. Leaves were put onto a smouldering fire, sometimes lightly covered with earth, and the sick person was made to lie or crouch in the warmth and aromatic smoke produced. Another medicinal is the corkwood ***witjinti***, the bark of which is burnt to ash and applied to sores resulting from burns.

MEDICINAL PLANTS

Trees	*witjinti*	corkwoods
Shrubs	*aratja*	native fuchsia
	irmangka-irmangka	native fuchsia
	karinga<u>n</u>a	mint bush
	munyu<u>n</u>(pa)	turpentine bush
	tjininy(pa)	hop bush
Subshrub	*ka<u>l</u>pipila*	—
Vine	*ipi-ipi*	caustic vine

Other plant uses

Plants of the region were, and still are, used by the Yankunytjatjara for purposes other than food and medicine. ***Mingkul̲(pa)*** is one of several species of wild chewing tobacco. The Yankunytjatjara highly prize the strong ***mingkul̲(pa)*** or ***ukir̲i***, known specifically as ***pulyan̲tu***, found in and around rocky situations. The dried, crushed leaves are mixed with the ashes of any of several species of plant, depending on availability. Most favoured is the bark of apar̲a, though when it is not close by, leaves of mulga ***kurku*** and sandhill grevillea ***yar̲an(pa)*** may be used.

Most traditional weapons, implements and utensils are made of the wood of a few different plant species. Mulga ***kurku*** is used to make boomerangs ***kal̲i***, spearheads ***wat̲a***, barbs ***mukul(pa)***, spear throwers ***miru***, digging sticks ***wana*** and fighting spears ***winta***. Hunting-spear shafts are obtained from the vine-like stems of the spear bush ***urtjan(pa)***.

The roots of the river red gum ***apar̲a*** provide the raw material for the variety of bowls, such as the ***mimpu***, principally used to carry water, the broader, shallower ***kan̲il(pa)*** used as a yandying dish and to carry foods such as fruits and witchetty grubs, and the cup-like ***wira***, used to collect berries and such, and as a digging tool. Adhesive gum ***kit̲i*** got from the leaves of the desert mulga ***minyur̲a*** or from spinifex grass ***tjanpi*** is used in jointing and affixing barbs to spears, setting quartz chips into the handles of spear throwers as an adzing tool and meat knife, and as a putty to plug small holes in bowls and other implements.

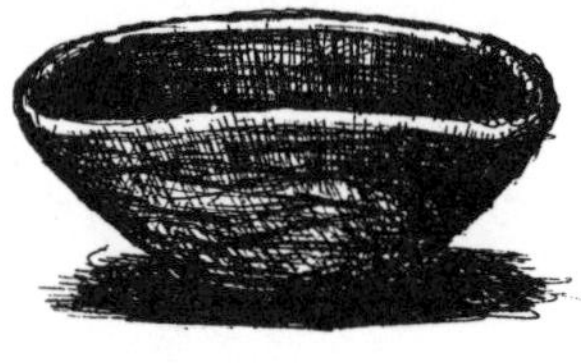

mimpu

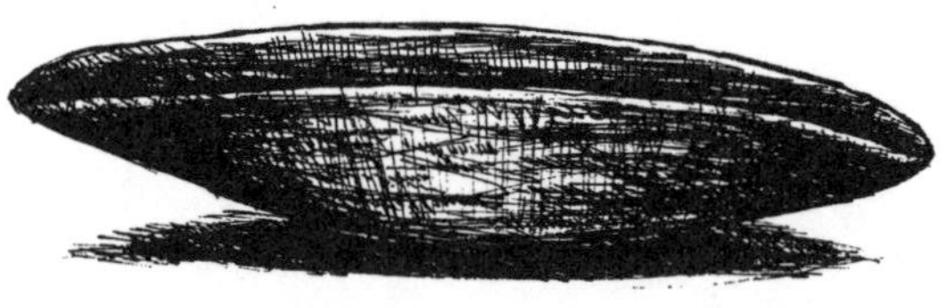

kan̲il(pa)

The emu poison bush ***tjila***, as the English common name suggests, is useful in drugging game. Crushed or pounded leaves are used to contaminate small rock pools from which the animals are known to drink. Once drugged, the stupefied animal is easy prey. Aboriginal people are generally concerned to point out that the ***tjila*** plant, also known as ***walkal(pa)***, must be handled with great care, and kept away from children.

Mulga ***kurku*** is the favoured plant used as firewood ***waru*** for heating and cooking, and also for the construction of windbreaks ***yuu*** and shade shelters ***kanku***, ***wiltja***. Other species, such as the river red gum ***apara***, are also used, depending on the location of a camp. Some trees, such as the silver-leaved water bush ***ilpara***, bear drinking water in their roots.

Plants are used in a variety of ways for personal decoration. ***Tjintjulu*** is the hair decoration made by threading strands of hair in small gumnuts ***tatu*** of the bloodwood ***itara***. ***Ilpatilpata***, a type of fungus known as a stalked puff-ball, is used as a body paint. The purple-black powder can be used by boys to make imitation beards, or by older men to darken white hairs in their beards. The sap of the caustic vine ***ipi-ipi*** is used to make decorative white spots on the skin. The kernel of the quandong ***mangata*** can be used to make a hair conditioner. Doubtless there are many other miscellaneous uses of plants known to the Yankunytjatjara.

There are many songs, stories and rituals of the creative period known as the Tjukurpa, or Dreaming, that relate to or involve plants. Increase rites ensure the persistence and availability of certain species. Characters of the Tjukurpa may be personifications of certain species. Though some is open, and available to all, much of the Tjukurpa lies in the realm of the secret/sacred ***miil-miil(pa)***, ***wiil-wiil***. It being better to err on the side of caution in such matters, the role of plants in the Tjukurpa and in ceremonial activities is not treated in this book.

The language of the stories

Yankunytjatjara is now mainly spoken in and around Mimili, Indulkana (Iwantja) and Fregon. There are speakers also at Ernabella (Pukatja), Amata, Mu<u>t</u>itjulu, Imanpa, Finke (Apatula), Kenmore Park (Yunyarinyi) and Oodnadatta (see the map on page 1). The Yankunytjatjara probably number some 300–400 people. Many Pitjantjatjara-speaking people also reside on these communities, and intermarriage is common. Although the traditional Pitjantjatjara homelands are to the west of Amata, the past hundred years has seen a substantial eastward movement of Pitjantjatjara people, so that many are now settled permanently within the traditional Yankunytjatjara region.

Communities like Mimili, where the contributors to this book have their home, are strongly traditionally oriented. The community language is Yankunytjatjara and/or Pitjantjatjara. English is learnt as a foreign language, and spoken only when necessary in interactions with non-Aboriginal community staff, teachers and visitors.

In linguistic terms, Yankunytjatjara and Pitjantjatjara are members of a much larger family of dialects, including also Ngaanyatjarra, Pintupi and others, called the Western Desert Language. Neighbouring dialects tend to be fairly similar, but across larger distances the differences in vocabulary, grammar and pronunciation build up to make mutual understanding more and more difficult. Nevertheless, the Aboriginal people of the vast Western Desert area share a common kin system, and are bound together in rituals and by affiliations to sacred places created by the ancestral beings of the Wapar, or Tjukur(pa), whose paths crisscross the region.

Speakers of Yankunytjatjara and Pitjantjatjara can understand each other very well, in spite of differences in common words, the forms of a few grammatical endings, and different accents. When words differ, older people know the counterparts in the other dialect, and sometimes use them. Younger people often merge or mix features of the traditional dialects. The table below summarises some of the important vocabulary differences (verbs are given in present tense form).

	Yankunytjatjara	Pitjantjatjara
come, go	yananyi	pitjanyi
travel, walk	yananyi	ananyi
return	kulpanyi	—
pick up, get	mananyi	mantjini

story, word, Law	wapar	tjukur(pa)
earth, land	manta	pana
rock, hill	apu	puli
far, distant	wanma	parari
middle-aged woman	kungka	minyma
sleep	anku	kunkun(pa)
true	mula	mulapa
all	kutjuli	uwankara
hungry	anymatjara	paltjatjiratja

Traditionally, many Yankunytjatjara words end in a consonant, where in Pitjantjatjara they would have the extra syllable ***-pa*** added. These days Yankunytjatjara speakers are adopting the ***-pa*** to some extent. You might, for instance, hear a Yankunytjatjara person say either ***malany*** or ***malanypa*** 'junior sibling', whereas a Pitjantjatjara person would almost always say ***malanypa***. In this book, words like this are written with the optional ***-pa*** in parentheses, e.g. ***urtjan(pa)*** 'spear bush', except in the stories, where they are written as actually pronounced by the Aboriginal contributor. Another difference is that few Pitjantjatjara words begin with a y, whereas many Yankunytjatjara words do — for instance, Yankunytjatjara ***yunganyi*** 'giving', ***yultu*** 'empty, hollow', ***yananyi*** 'travelling, coming, going' correspond to Pitjantjatjara ***unganyi***, ***ultu***, and ***ananyi***.

The language found in this book is probably closer to the traditional or classical Yankunytjatjara than that in common use about camp, because the contributors, all of whom are older, senior people, were generally making a conscious effort to achieve this.

Vocabulary

Like other Aboriginal languages, Yankunytjatjara has a rich vocabulary, probably containing around 10,000 words. It includes an extensive botanical vocabulary, such as plant names (see section 4.1) and terms for stages of plant growth and parts of plants (see section 3.1). There are also many words used to describe the processes involved in preparing plants for use (see section 3.3). Some of these are quite specialised and require a phrase of several words to be fully translated into English. We have only space here to mention the most common of the unusual features of vocabulary from an English speaker's point of view.

Yankunytjatjara does not have separate pronouns for 'he', 'she' and 'it', all of which are covered by the word *paluṟu*, or a variant of this. Rather than write 'he or she' over and over in the translations, I have taken the approach of choosing either 'he' or 'she' depending on the context. This means that 'she' occurs more often than 'he' because plant foods are mostly gathered and prepared by women.

Perhaps naturally for a language suited to the needs of a highly mobile culture, Yankunytjatjara has quite a rich set of common expressions referring to directions, orientation and relative distance. For instance, there are four words that can be translated as 'this' or 'that', depending on whether something is physically present or not, and on how distant it is. Words like this — linguists call them demonstratives — are used much more commonly in Yankunytjatjara than in English. To make the translations more natural in English these are often not translated to their full accuracy, or are rendered simply with the English 'the'.

Often, simple Yankunytjatjara words require different translations depending on the context. The word *puṉu*, the title of this book, is a case in point. It can mean 'tree', 'bush', 'plant', 'wood'. Another important example is *mai*, which can mean 'vegetable food' or 'food plant' — i.e. product and source are covered by the same word. As in most Aboriginal languages, there is no general word for 'food': as mentioned earlier, Yankunytjatjara distinguishes vegetable food *mai* from meat/flesh food *kuka* and edible grubs *maku*. Another interesting Yankunytjatjara word which has to be translated according to context is *nyinanyi*, which may mean 'sitting', 'living' or 'staying', and sometimes just 'is'.

As in many languages, a good number of the terms for the parts of a plant are extended or metaphorical uses of words for parts of the human body — for example, the branch of a tree is an 'arm' *miṉa* and the stems and trunk are the 'body' *aṉangu*.

Some words are so culture-specific that it is inconvenient to give a full explanation of them each time they come up. The most obvious examples are words for multi-purpose implements and tools. A *wira*, for instance, is like a small wooden cup used for drinking, scooping out earth while digging, and for collecting berries. The main use of the *miru* is as a spear thrower but it also serves as a meat knife and adzing tool through the quartz chip set into the handle. Probably no single word or short phrase in English can convey the full concept behind these words.

Other problem words for the translator involve spiritual or religious beliefs. Aboriginal people believe there are various kinds of dangerous, invisible spirit-monsters ***mamu***, which can bite and enter the body. These are a possible source of illness and pain that can only be treated by an experienced healer-magician ***ngangkaṟi***. Again, no short English phrase can capture the meaning of these words. In this book, words like ***wira*** and ***mamu*** are carried over into the translations, sometimes as part of a larger phrase e.g. a ***wira*** dish, a dangerous ***mamu*** being.

Grammar

Yankunytjatjara has an elegant grammatical structure, comparable in some ways with that of Latin. For more information, consult the IAD Press publications *A Learner's Guide to Yankunytjatjara* and *Yankunytjatjara Grammar*, and the work on Pitjantjatjara listed in the References section. As in Latin, the role that people or things play in the situation is indicated mainly by case endings attached to words. This means that word order is much more flexible than in English (though usually the verb comes at or toward the end of a sentence).

The relative time that actions are done is indicated by tense endings that attach to verbs. There are in fact four slightly different sets of these endings, depending on which one of four groups or 'verb classes' a verb belongs to. Verbs may be in the present, future, or past tenses (continuous or non-continuous); or they may be imperative/potential (continuous or non-continuous). There is also a special 'characteristic' verb ending, used to refer to actions as typically or usually done by someone. The stories in this book mostly use the present tense or the characteristic ending. The present tense is used to list verbs in the Glossary (Part 3), and in most dictionaries and wordlists.

Yankunytjatjara grammar also has some features that are foreign to European languages, but quite usual in Australian languages. The first we will mention here is what is called the serial verb form. This allows a number of verbs, usually depicting actions carried out one after another or together, to be telescoped into a single sentence. For instance, in ***ngayulu yankula nyangu*** 'I went (and) saw it' there is no Yankunytjatjara word corresponding to the English 'and' — the word for 'go/went' ***yankula*** is simply said in its serial form.

Serial verbs are extremely common in Yankunytjatjara, especially when, as in the stories told in this book, speakers are going over a series of interconnected actions, such as gathering, cleaning, grinding and baking seeds.

Skilled storytellers are fond of using serial verbs to recapitulate the preceding steps in the process before each new one is introduced. To get a similar effect in English, it is often necessary to use a phrase beginning with a word like 'after' or 'having'. For instance, ***mai palunya kaṉiṟa, rungkaṟa, pauṉi*** might sometimes be translated as 'after yandying and grinding the seed, you bake it'.

Repetition of the same serial verb is used to indicate that something is done over and over again, or kept up for an extended period of time. Sometimes this is best translated into English using a phrase like 'for a while' or 'over and over' or 'more and more'. It is quite common to hear from two to six repetitions of a verb. To save space, these are indicated in writing by a figure in parentheses. ***Rungkaṟa***(4), for instance, indicates the word ***rungkaṟa*** 'grinding (serial form)' repeated four times.

The Yankunytjatjara examples so far illustrate a second unusual feature of the language, from an English speaker's point of view. Often when the identity of a person or thing being referred to is obvious in context, a person speaking Yankunytjatjara will make no mention of it. From an English point of view this means that words like 'it', 'something', 'he' and 'she' are often left out, giving Yankunytjatjara speech a compact, abbreviated quality. This is very difficult to convey in translation — in English, for instance, it is ungrammatical to say 'I saw'; so to translate ***ngayulu nyangu*** we must choose between versions like 'I saw it' and 'I saw something'. Readers studying the Yankunytjatjara stories should be alert for this.

In English, probably the most natural way to talk about how to do something, without having any particular person in mind as the actor, involves using the word 'you' in a special non-specific way, e.g. 'you gather quandongs in a ***wira*** dish'. Yankunytjatjara gets the same effect by not mentioning any particular actor, e.g. ***mangaṯa wirangka uraṉi*** (literally) 'quandongs in-a-dish gathers'. Perhaps the most literal translation in English would be 'one gathers quandongs in a ***wira*** dish'. But because frequent use of 'one' in this way makes English stories sound rather highfalutin, the non-specific 'you' is generally used in the translations in this book.

Aside from the obligatory endings mentioned above, Yankunytjatjara makes extensive use of additional, optional endings such as ***-lta***, ***-mpa*** and ***-nti***; generally these may attach to any word in a sentence, and convey nuances, indicate the speaker's attitude to something being said, and so on. Very roughly, ***-lta*** marks a significant turning point in the events being spoken about, ***-mpa*** focuses interest on what will come next, and ***-nti*** suggests doubt. There are many other optional endings (clitic particles), but these are the most common in the stories in this book.

Finally, a quick glance at the stories will show that there are two sentence-linking words ***munu*** and ***ka***, one of which is found at the onset of almost every new sentence in a connected story. To some extent, it is fair to say that these could both be translated as 'and' (and ***ka*** also sometimes as 'but'); except that ***munu*** is confined to joining sentences that share a single subject or actor, and ***ka*** generally joins sentences with different subjects. On the other hand, both ***ka*** and ***munu*** are used much more frequently than 'and' is used in English: in a way they are like oral punctuation. Often, therefore, they are not translated at all in the English versions of the stories.

Sound system

The spelling system used in South Australia to represent the significant sounds of the language was devised by the Ernabella missionaries some forty years ago. Slightly different systems are in use in Western Australia and the Northern Territory. The South Australian system uses a selection of letters from the English alphabet, and an extra mark or 'diacritic' — the underlining which can be used beneath some letters.

The consonants in the Yankunytjatjara and Pitjantjatjara 'alphabet' can be set out as follows, grouped according to how they are pronounced. Sometimes a combination of two letters is used to write a single sound, just as in English **sh**, **th** and **ph** each represent only a single sound.

stops	p	tj	t	ṯ	k
nasals	m	ny	n	ṉ	ng
l-sounds		ly	l	ḻ	
other	w	y	r	ṟ	

Unlike English, Yankunuytjatjara has only one set of 'stop' sounds. In English there are two, distinguished by 'voicing' — whether or not the vocal cords are vibrating as the sound is made in the mouth. As far as the mouth is concerned, the English sounds written **p** and **b**, for instance, are often pronounced in virtually the same way — the difference is whether or not the vocal cords are involved. English **t** and **d** and **k** and **g** differ in the same way. When learning to speak English, a child or adult learns, consciously or unconsciously, to pay attention to this difference, and to control it while speaking.

In Yankunytjatjara, voicing is not used to distinguish any pairs of sounds.

This means that from an English speaker's point of view, the sounds in Yankunytjatjara that are spelt ***p***, ***t***, and ***k*** are often pronounced like **b**, **d**, and **g**. The point is that because the difference is not important to distinguishing meaning in Yankunytjatjara, the pronunciation can vary, often due to the influence of the neighbouring sounds in a word.

There are two sets of Yankunytjatjara sounds pronounced in a way quite foreign to English. These are the retroflex sounds ***n̲***, ***t̲***, and ***l̲***, and the 'teeth' sounds ***tj***, ***ny*** and ***ly***. Retroflex sounds, as in ***pun̲u*** 'plants, wood', ***mal̲u*** 'kangaroo' and ***tjut̲a*** 'many, plural' are done by curling the tip of the tongue a little further back in the mouth. This gives a thicker sound, with an **r**-like quality, something like the way most Americans pronounce the words 'corner', 'surely' and 'warder'.

With the teeth sounds (laminodentals) the front part of the tongue is thrust forward so it touches the backs of both sets of teeth. You can find the position by putting the tip of the tongue at the base of the lower teeth, and pushing the tongue forward. ***Ny*** and ***ly***, as in ***nyuntu*** 'you' and ***palya*** 'good', sound a bit like the English sounds in the words 'onion' and 'million'. ***Tj*** as in ***tjut̲a*** 'many, plural' sounds a bit like **ch**, as in 'church'.

There are two **r**-sounds in Yankunytjatjara. The 'smooth' ***r̲***, as in ***war̲u*** 'fire, firewood' is like the ordinary English **r**-sound, as in 'road' or 'arrow'. The 'sharp' ***r*** is either flapped or rolled, as in some European languages and Scottish English.

Many Yankunytjatjara words begin with the sound ***ng***, which in English is found only in the middle and at the end of words, e.g. 'singing'. To learn to say Yankunytjatjara words like ***ngayulu*** 'I', ***ngura*** 'camp, home' and ***ngunytju*** 'mother', it helps to bear in mind that you don't have to do anything new or difficult with the tongue. The whole mouth behaves exactly as if you were saying a **k** or **g**. The difference is in fact whether the nasal chamber is involved, making ***ng*** a so-called 'nasal' sound.

Often English speakers can learn to say ***ng*** at the beginning of a word in the following way. First, practise and concentrate on how to say ***kaa*** (car). You will notice that the back part of the tongue humps up briefly to touch somewhere in the rear of the upper mouth. Now get the tongue and mouth ready to say ***kaa*** — actually position the tongue in this familiar position, relax and just try to produce a ***ngaa*** sound. Once you stop thinking that any funny tongue movements are involved, the nasal quality comes easily.

There are three significant vowel sounds in Yankunytjatjara, written with the letters ***a***, ***i*** and ***u***; ***a*** usually sounds as in 'father' but a bit shorter; ***i*** and ***u*** sound as in 'pizza' and 'put'. The first vowel in some words is long. This can be written by using two letters, as in ***tjaa*** 'mouth, language', or with a colon following the vowel, e.g. ***tja:***. Some Pitjantjatjara reading material does not write long vowels any differently than short. In this work the first system is used, e.g. ***nyaa*** 'what?', ***nyii-nyii*** 'zebra finch'. Some combinations of vowels occur, as in the words ***mai*** 'vegetable food, food plant' and ***pau<u>n</u>i*** 'bake, roast, parch'. (In the Northern Territory and Western Australian spelling system there would be a ***y*** or a ***w*** separating the vowels, as in ***mayi*** and ***pawu<u>n</u>i***.)

Because there are only three types of vowel, each can vary quite a bit in pronunciation, depending on the neighbouring sounds in a word. In words like ***wangka*** 'speech', ***wala*** 'quickly' and ***apungka*** 'on the rock, in the hills' the ***a*** usually sounds like an **o** — because the rounding of the lips of the previous sound is carried over to affect the ***a***. When an ***i*** comes before a sharp ***r***, as in ***miru*** 'spear thrower' or ***wira*** 'digging dish/cup' it usually sounds more like an **e**.

Yankunytjatjara words are almost always pronounced with stress on the first syllable. In fluent speech, this is accentuated by a number of contractions, which have the effect of running words together, so that speech

Ara<u>r</u>anganyi: *winnowing seeds from chaff*

sounds rhythmic and flowing.

The verbal culture of the Yankunytjatjara is elaborate and diverse. Aside from a huge body of oral literature comprising Dreaming stories, there are highly developed styles of yarning, joking and oratory. With this background it is little wonder that most older Yankunytjatjara people are skilled storytellers, able to make even short texts, like most of those in this book, charming and entertaining as well as instructive.

YANKUNYTJATJARA PLANT USE

Trees

apaṟa **river red gum**

iṯaṟa *Eucalyptus camaldulensis* var. *obtusa*

A white, sugary scale insect secretion ***ngapaṟi***, ***apaṟuma*** that forms a flaky crust on the leaves is a favoured food of adults and children. Branches are collected and put on a hard surface to dry the ***ngapaṟi*** off. After being made into balls, it can be eaten at leisure. Edible grubs ***itjaḻiti*** are found in the trunks and branches, and those called ***unganungu*** in the roots. The bark ***likara*** is burnt to ash ***unu*** and mixed with chewing tobacco or 'pituri' ***mingkuḻ(pa)***, ***pulyaṉtu***. The roots ***iwiṟi*** are used to make bowls ***mimpu***, ***kaṉil(pa)***, ***wira***, ***tjakun(pa)***.

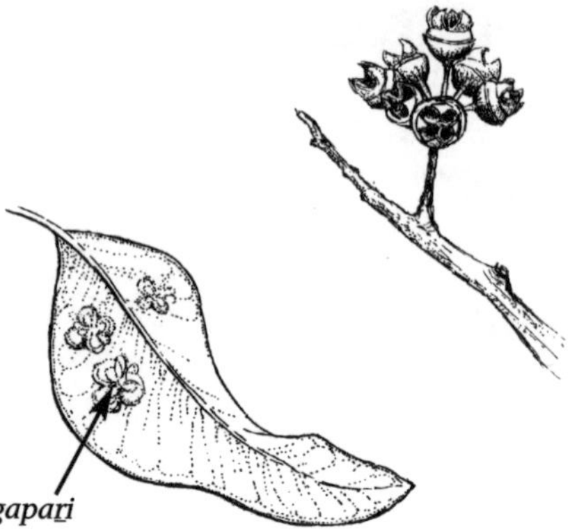

A magnificent, tall spreading tree to 35 m high — one of the largest in the region. Smooth, variable-coloured bark, often green; buds, flowers and fruits in clusters; flowers cream; fruit a small capsule with four valves.

A common tree in alluvial soils of watercourses and floodplains, often forming woodlands.

Pompey Everard

Ngapaṟi kungkangku watingku kuriṟara kuriṟara wampanti — pararitjangku pararitjangku kapuṯunkupai, uraṟa.	Women, men, husbands and wives, anyone on a trip, would make it into a ball, as they gathered it.
Ka katira(2) kapuṯuṟa(9) puḻka kutu kapuṯunkupai.	They used to carry it along with them, making it bigger and bigger and bigger — into a really big ball.
Kapuṯu puḻkanya katira tjuṯa yungkupai, tjitji tjuṯa, wati, kungka, wati, kungka, tjitji.	They'd take that big ball and share it out — among the children, the men and the women.
Wampanti ngalkupai, walytjangku waḻi-waḻi. Munu tjunkulanku puṟinymaṟa ngalkupai.	Everyone has some, oneself as well. After putting it down and softening it, you eat it.

mangaṯa **quandong**

wayanu̲ ***Santalum acuminatum***

The skin and flesh of the fruit is an important and much relished food ***mai pulka***. The dried fruit ***mulya*** may be reconstituted with water and eaten. The stone ***taṯu,*** cracked open, reveals the kernel ***kuuti***, which is mashed to an oily paste and used medicinally for aches and pains. The wood is used to carve snakes, lizards and similar items for sale and the leaves are used for smoothing these down.

A medium-sized tree to about 5 m tall with a dense, distinctive green canopy. Leaves somewhat thickened, opposite, tapering to a point; white to cream flowers borne in sprays on the end of branchlets; fruit when mature has a red skin, yellow flesh and a deeply pitted spherical stone, with an oily kernel; a root parasite.

Occurs occasionally in the region, mainly in sandhill and sandplain country. Flowering time is spring and summer. Fruiting in winter, spring and summer.

Pompey Everard

Ka kutjupangku watjalpai 'Wanyula mangaṯaku yara!'	Someone would say, 'Let's go for quandongs!'
'Kungka tjuṯa, pakala! Kala mangaṯaku yara!'	'Women, come on! Let's go get some quandong!'
Kungka tjuṯa yalṯi-yalṯipai, tjananku,	The women would all be singing out

kungka tjutangku. Wati wiyangku wantipai.	like this, to each other. They wouldn't call the men.
Munu yankula nyakupai. 'Palatja purkupupanyi, purkupupanyi.'	Then they'd go find a tree. 'Oh, there's one laden with fruit — it's borne down with fruit.'
Ka urara(9), mulya tjara kutjupa uranma. Mulya urara(2).	They'd gather and gather plenty of it, and they'd also gather some of the dried fallen fruit.
Pilti nyanga urara(3). Mai wiru tjuwita kutu.	The dried-out stuff is really nice and sweet.
Arkara urara(3) pulaparira katipai.	After tasting it, they'd gather it till it filled their bowls, and take it away with them.
Munu katira ngurangkalta kutjupa tjara paulpai. Paura, tjukutjuku nguwan paulpai.	After bringing it back to camp, they'd bake some of it. Just bake it a little.
Munu palunya urara kapingka kilinara paltjira, kilinarampa, alatji, alatji katantanama, munu tjunama(2). Katantara tjunkula(3).	Then get it out and clean it in water — wash it, and when it's clean, break the flesh off like this, and put them to one side. Break them off and put them to one side.
Mai pii panya. Mai palunya katantananyi — lampiralta tjunkula(4) alatjingara.	The edible flesh of the plant. They'd break off the flesh, peel it off and put it to one side.
'Wanyuna kaputura tjura!' Kapukaputura ma-yungkupai	'I'll just make this up into balls!' She'd make it into little balls and give it out.
Munu kaputura tjunama(2) munu watjalpai:	But she puts some to one side, and says:
'Tjunyinyina kuwarimpa, ngayulumpa. Uwa, ngayulu tjunyini.'	'I'm going to soak this, directly, I am. Yes, I'm going to soak it'.
Tjunyiralta kapi pulkangka, nyara palunya tjunyini.	And she'd soak it, soak it in plenty of water.
Tjunyira(10) wiruralta ma-tjunkupai.	Squish it around for a long time, really thoroughly, and then put it to one side.
Ka tatulta kilinara(7). Uwa nyanga kuuntjara palku ngalkunma.	The kids would clean off the stones. Yes, by sort of sucking off the remaining flesh.

Mai wali-wali ngalkunma. Ka 'Nyuntu mai ngalkula wiyalku, tjitji awa!'

They might try to eat some of the food that's been soaked. 'You'll eat it all up, you kids!'

'Tatu palatja kutjuya kilinanama. Tatu kutju. Mai munkaratjara ngalkuwiyangku.'

'Just clean the stones, only the stones! Don't eat what's been put aside!'

Kungka ngunytjungku pailpai, ngalkula(2) tjitjingku.

The mother would shoo the children away as they ate.

Ka tjitjingku kilinara wantipai. Wati panya mamakulta wantipai.

And so the kids cleaned the stones, and left the rest alone. They wouldn't touch what was left for their father.

Ka mamangku, wati mamangku kulpara nyakupai, kungka panya.

And the father would come back and see what the woman had done.

'Ngayuku kuringku nyangatja tjunyira wirura tjunangi mangata.' Alatjingara paluru ngalkupailta.

'My wife here has soaked and made up some quandong!' And then he'd eat.

Ngalkula(7) ma-wantipai. 'Awar!' Talturingkula.

And after eating and eating, he'd leave off, having got full. 'Oh, yes!'

Ka tjunyira. Kutjupa ngarimatu. Mai pulka kutu ngarima,

There'd still be plenty of the soaked stuff left. There'd be lots left.

ka kutjupangku(3) katjangku(3) wati pakalingku maitu ngalkunma, mai palunya. Mai pulka nyangatja.

So all the others — sons and grandsons, would have some too. It's an abundant food.

Munu 'Nyangatja kaputuna wai?'

'What'll I do with this ball of quandong?'

'Uwa, yuwani.' Katira paluru ngurangka ngalkuntjaku.

'Yes, give it to me!' He'd take it back with him to eat at camp.

Ngarira, nyinara(2) paku alira, nyinara(2) kulira, ngalkupailta.

Then he'd have a sleep, rest a while, and when he felt hungry, he'd eat.

Uu kaputu panya paluru, wirunya, kuka wiyangka.

Yes, from that same ball of quandong. It's really good, when you don't have meat.

Mollie Everard

Mai katutja, ngananana urara(2) katira ngalkupai,	The edible part on the outside, we'd gather and eat.
munula tatu — tatu ngapartji utulura atura, lampalpungkula urara(3) — kuuti panya unngutja.	And as for the stones — we'd collect them too, hit them with a rock, prise them open, and get the kernels — you know, the inside part.
Urara(7), utulura ka ngayuku kamingku utulura rungkalpailta.	We'd gather up a whole lot of them. My grandmother would gather them up and grind them.
Wirangkalta rungkalpai, atura wirura.	She'd grind them in a wira dish, after chopping them up well.
'Pika nyitintjitjangku.' Pika nyitintjitjangku atulpai. Nyanga unngutja kuuti panyampa.	'It's to rub into aches and pains.' You know, the kernel inside.
Paluru tjana mai nyanga pii katutjampa nyaampa, mai ngalkuntjampa, paluru nyara.	The skin on the outside they'd eat as food.
Ka unngutja, kuuti nyara, raapamilalpai witapilta. Witapi pika ngarala, raapamilalpailta.	And they would rub the kernels into the back. If you had a backache, you'd rub it in.
Ngarira tjinturingkula pakara kulilpailta: 'Ai! Pika panya ngayulu wiyaringu!'	Sleep the night, then get up and think: 'Hey! That pain of mine has gone!
'Ngaltutjara! Miritjina panyangka ngayulu palyaringu.'	'The dear thing! With the aid of that medicine, I've recovered.'
Munu ngura palyaringkula kukakulta yankupai.	They'd just get better and go off for game.
Uwa, nyanga palunya kutjuna wangkanyi.	Yes, that's all I'm saying.

minyura — desert mulga

kurku — *Acacia minyura*

A white powder encrusting the leaves is a source of resin ***kiti*** used in joining spear parts, setting quartz flakes in handles of spear throwers and adzing tools, and repairing cracks or plugging holes in wooden bowls.

A rounded, shrubby form of mulga to about 4 m high with small, distinctive, curved-elliptical, bluish leaves. The yellow flowers are in cylindrical spikes; pods are small and flat.

Much less frequent than other common forms of mulga, occurring mainly in the loamy sands of plains, and in the deep red soils of undulating sandplains and sandhills.

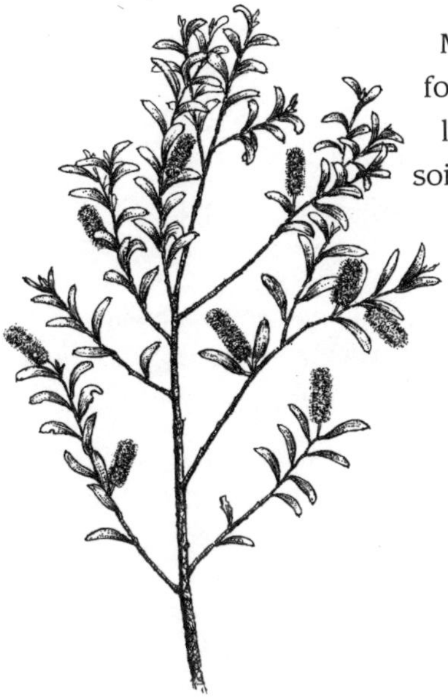

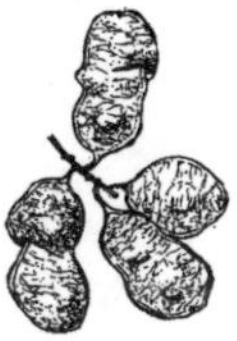

Kanytji

Ka ma-nyakukatira putingka nyakukatira(3) nyakupai, ilurangkula nyangatja ngaranytjala. Uti kutu.	Keeping a look out as you go along — as you travel through the scrub country you see the desert mulga standing out clearly. Really plainly.
Ilurangkula ngaranyi, tiirangkula panya, tiin puriny.	They stand out, because they sort of shine, like tin.
Ka marangkulta muurpungkula, katantara alatjingara...	You crush up some leaves in your hands, breaking it up like this...

Ular-ularpungkula nyakula, 'Ai, marangka puyilarani, kiti nyanga pulka.'

As you rub it you see, 'Ah, it's getting greasy on my hands, there's a lot of resin here.'

Munu katantaralta tjunanyi, munu tjunkula(3), tjunkulampa putulta para-nyakula.

Then you start breaking some off, and once you've got a whole lot you look around for an old flattened termite nest.

Kilinara(2) mula, putjangka munira(4) kilinara mulalta, tjuultjura(5) pulkara kutulta kutjuwara kutulta pungkula(3) punungka pungkula.

You clean off that hard ground well, sweeping it with grass — clean off all the loose dirt, then, after piling a whole lot of leaves on, to get it done in one go, you thresh it, pound it with a stick.

Ka kampa kutjupara pungkula(4) kampa kutjupa pungkula(2), wiyarampa, utulura nyanganyi.

Then turn it over and pound it some more, pound the other side. Then when it's finished you gather it together and have a look.

'Au, ngayukur pulka kutu, pulaparingu!'

'Oh, there's a nice big lot there for me!'

Ka ngura utuluralta, parkangka ilara(5) ilarampa, warulta kutjani.

And you just gather it up, pulling it together with some leaves, and light a fire.

Munu kutjara likara tjaatjura witara(3) munu punungka yuritjinganilta. Nyanga purinymananyi.

After lighting a fire, you light the end of a piece of bark and lightly heat the resin powder, and move it around on a twig, like this.

Marangka witi-witira kaputura(3) piyuku kuti-kutini, nyanga palula.

By sort of grasping it lightly you make it into a ball, and twirl it round and round again, on the same stick.

***Yalkani**: threshing the desert mulga to separate the resin*

Kuṯi-kuṯira(2) kuṯi-kuṯirampa... pulkaringkunytjitjangka kulpalyinanyilta.	You twirl the stick around and around and around, and once it's got really large, you take it home with you.
Mantangka panya kilinaṟa mulalta, manta panya, wiṟura, kutungku, kutjuli uraṟa mulalta.	After cleaning it up off the ground really well — only after getting the whole lot up.
Munu ngura kulpara panyalta tjunama.	Then you just bring it back to camp with you.
Nyangatjampa waṟi wiyangkalta miru nyangangka atuṉi, kiṯitjarangku kuliṟa.	Once you've got that you can make a spear thrower without any worries, knowing you've got some resin.
Munu katji, katji kaṯantananyi, kiṯitjarangku kuliṟa, palkatjuṟa waṯatjuṟampa ngura kuka wakaṉilta.	And you can go break off some spearvines, knowing you've got enough resin. Once you've put the blade on the spear and the endpiece, you can just go spear game.
Iniwai kutulta, alatji...Alatji kutjuka.	Any way you like. That's it. That's all there is to it.

wintalyka — mulga

kurku *Acacia aneura*

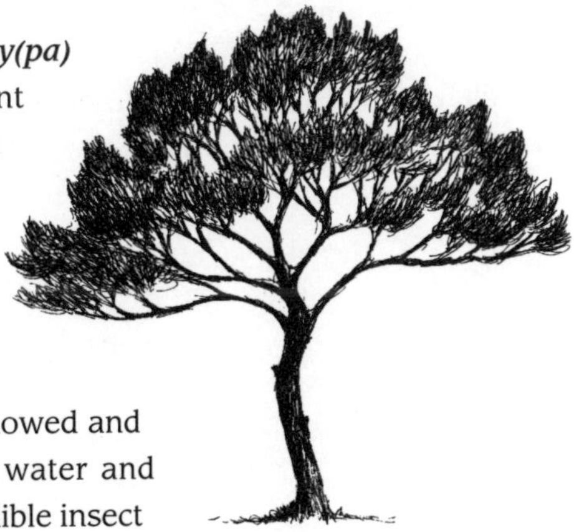

The mature, hard seeds **kalka, uṉiny(pa)** are one of the most important plant food sources. The pods are collected and the seeds are separated from them by threshing and rubbing. The seeds are then yandied to separate them from the remaining pod pieces, and the clean seeds are parched in hot sand and ashes, winnowed and yandied, and then moistened with water and ground to an edible paste **latja**. An edible insect gall ***tarulka***, the size of a large marble (2–3 cm in diameter), is found on branchlets at certain times; it has a slightly reddish skin when ripe.

A clear, sweet exudation ***kurku***, ***kurkun-ytjungu*** drips from upper stems of branchlets when sap-sucking scale insects, which form reddish scales, are active. This 'honeydew' ***tjuratja***, ***wama*** can be sucked directly from plants, or it may be dissolved in water to make a sweet drink. When dried out, ***kurku*** forms reddish lumps known as ***mangir(pa)***. The mulga apple ***tarulka*** is an edible insect gall, eaten raw. The hard wood is used to make spear throwers ***miru***, barbs ***mukul(pa)*** and spearheads ***wata*** for spears, boomerangs ***kali*** and digging sticks ***wana***. It is also excellent and abundant firewood ***waru***.

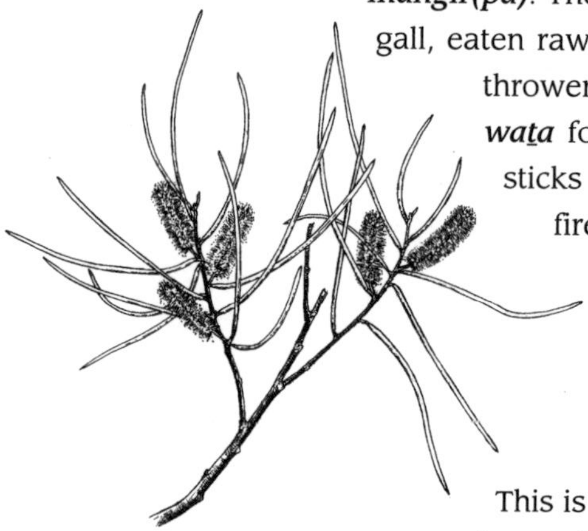

This is the most common form of mulga in the region and varies considerably in appearance: it may be shrubby, with several trunks, or a small tree with a single trunk. Leaves are extremely variable, though often needle-like, slightly flattened to about 5 cm long; flowers in cylindrical heads to 3 cm long; pods brown, papery and flattened, with a few ovoid seeds in each.

Often occurs as a dominant species on the loamy sands of plains, on rocky slopes, and in the sandy soils of flat or undulating sandplains and sandhills. Extensive mulga groves may be found in some areas. Within one area several different forms of mulga may occur. Fruiting usually in spring and summer.

Puṉu putungka tjunkupai. Putungka tjunanyi munu wantinyi.	You put some mulga branches on an old flat termite nest and leave them there.
Munu nyinara(2) yankula(2) nyanganyi. Pilṯiringanyi.	Then after a while, after travelling around a bit, come back and check it. It dries out.
Ka 'Ai! Ngayuku mai pilṯiringu!' Panya palunya yalkaṉilta.	'Ah! This food of mine has dried out nicely!' Then you thresh it.
Yalkaṟa ka purputukatinyi.	As you thresh it, the pods fall out all over the place.
Ka nyulkuṟa palunya pungku-pungkula kaḻkani, munu palunya wirangka tjunanyi.	By rubbing them and tapping them you get the seed out, and put it in a **wira** dish.
Munu paluṟu kaṉini, kaṉini palunya. Wirangka kaṉini, wintalyka palunya.	Then you yandy the mulga seed in the **wira** dish.
Kaṉini, munu tjaṟuwaṉinyi, mimpungka tjaṟuwaṉinyi(2).	Once it's yandied, you tip the clean seed into **mimpu** bowls.
Munu palunya tjaḻiṟa kulpanyilta.	Then load up on your head and return to camp.
Munu ngurangka paluṟu kaṯira tjunkula, ipangka pauṉi, mai palunya, ka paluṟu pilta-piltatjunanyi.	Having brought it to camp, you parch the seeds in hot ashes, and the hard seed cases crack open.
Ka paluṟu pauṟa, araṟangkula, kaṉinilta. Kaṉiṟa kaḻkani, mai palunya.	After roasting it, you winnow and yandy it again, to separate the seed inside.
Tjiwa uraṉi, munu palunya rungkaṉi, rungkaṟa anytjuṉi.	Then you get a grindstone, and grind it up, licking it as you grind.
Munu nyaangka, tapaltjuṟa palunya mai rungkaṉi, mai palunya. Ka paluṟu tapalta pulaparingu.	You put a collecting vessel beneath the lower grindstone and it builds up in that.
Ka ngura tjitji tjuṯangku ngananạ ngalkupai. Ngananạ talṯuringkupai.	We used to eat it as kids. We'd get full.
Mai paluṟu puḻka, tjamuku, kamiku, mai kuṉpu. Ngananạ mai palula puḻkaringu, palawa wiyangka.	It's an important food: a food of our grandfathers and grandmothers, a strong food. We grew up on this food, without flour.

Mollie Everard

Nyakukatipai. Ka 'Tjulpu nyara nyawa! Tjulpu nyara ilura waninyi.'

You spy it as you're on the move. 'Look at the birds over there! They're all over the place.'

Pika ngurpatja nyakukatipai, pirpi-pirpira wanani.

You see — oh, just too much honeydew for words, glittering along the branchlets.

Ka ngura munira(3), katantara(6) munira(2), tjunkula munira(3) piyuku yanama, kurku kutjupakutu.

Well, you just strip the branchlets of leaves, and break off a whole lot. You strip them off, put them down, and then go to another mulga.

Munu katantara(2) ngalya-kulpama, munira(3). Piyuku kurku kutjupakutu yanama.

Break some off and come back, after stripping the leaves. Then you go once more to another mulga tree.

Munu katara tjunkula munira(3) kurku kutjupakutu, kutjupa katara tjunkupai.

Break some more off, strip off the leaves, then to another tree, for some more.

Munira munirampa kunakarpungkupailta.

After stripping off the lower branches, you chop off the upper foliage, after laying the branches side by side on the ground.

Tjunkula wanara, tjunkula wanara, kunakarpungkula wanara.

After they've lined the branchlets up side by side, they chop the upper foliage off all at once.

Ngalya-utulura wanara, putjangka tjunguralta, karpira tjunkupai. Karpira tjunkupai. Kutjupa kutjupa tjunkupai.

Then they gather them together into bundles, tie them up with grass and put them aside. They make one bundle after another.

Kutjupangku kutjupangku kungka tjutangku, tjunkula(4) tjunkulampa tjaliralta katipai nguraku.

All the different women, a whole lot of women — after they make lots and lots of bundles, they load them on their heads and take them to camp.

Katira ma-tjarpapai ngurangka,

As they arrive at camp, they'd say:

'Ngurangkala kapi tjutila, kala puta tjunyilalta, kurku nyanga!'

'Let's get some water from camp, and then what d'you say we go soak this mulga!'

Tjutira(3) katira, mimpungka tjunkula(4) tjunyilpailta warkungka, warku warungka tjunyilpai.

They get plenty of water in ***mimpu*** bowls, and then go soak the mulga honey in water, in shallow little depressions in the rocks.

Kampanytjala tjintungku kampara waruntjala apu.

Where it's really hot because the sun's heated up the rock.

Tjunyira(3) waru kutultala tjikilpai, kupalku tjikilpai. Apu warungka tjikilpai. Tjunyilpai nyangatja, kurku.

After soaking it a while we'd drink it really hot, as a hot drink. From the hot rock. You soak this stuff all right, mulga honey.

Tjutira(3), kungkangku katipai nguraku.

Then the women get it out and take it back to camp.

Ka tjitji tjutangku tjikira(3) palulaka kaltarapungkupai.

And when the kids drink it, it makes them burp.

Palunyana wangkangu kutju.

That's all I've got to say.

kurku or **kurkunytjungu**

witjinti **corkwood**

Hakea lorea ssp. *lorea*
Hakea divaricata

Corkwood flowers ***ul̲ar̲ama*** are a favourite source of nectar ***tjuratja, wama***. They are sucked directly, or gathered and squeezed in water to produce a sweet drink. The bark ***likara*** is burnt to ash and applied to burns.

Hakea lorea ssp. *lorea*

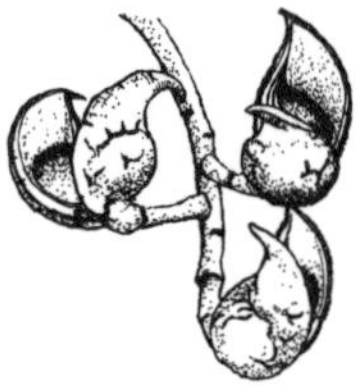

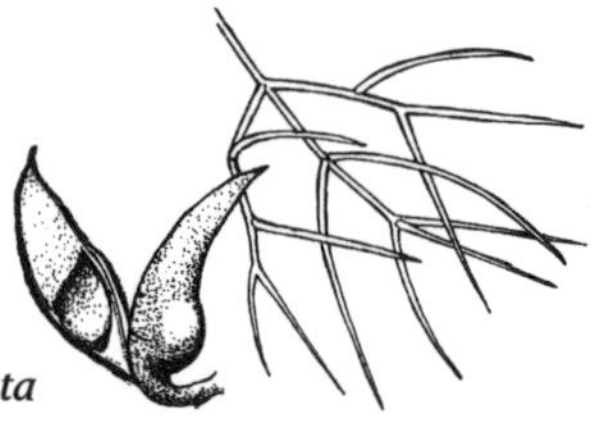

Hakea divaricata

Hakea lorea ssp. *lorea* is a small gnarled tree, 4–5 m high. The bark is thick, corky and furrowed; leaves cylindrical to 30 cm long, with a sharp point; flowers borne in showy yellow clusters; clustered fruits are woody. These capsules ***mar̲a*** are beaked, and contain two winged seeds which are released when the capsule splits open. *Hakea divaricata* differs primarily from *Hakea lorea* ssp. *lorea* in having shorter, forked leaves.

Occurs in sandy loam on plains and at the base of ranges, and in rockier soils on the footslopes of outcrops, hills and ranges, usually singly, or forming low, open woodlands. Flowers in winter and spring.

Pompey Everard

Ka witjinti tjunyilpai.	You soak the corkwood flowers.
Witjinti palunya katantara tjunkula(3), kapingka tjalatjura(3), kapingka tjalatjura(5), kapi pulkangka.	You break off a pile of flowers, then immerse them in water, putting in more and more — in a lot of water.
Tjalatjura(3) palunya panya tjunyinma.	More and more goes in the water, and you soak it.
Munu piruku malatja katantara tjunama.	And afterwards break off some more to put in.
Katantara kapingka tjunyira(2) ka kapi paluru ma-maruringulta.	After you break some off and soak it, after a while the water goes dark.
Ka kutjupalu, 'Ayi, tjinguru wama, tjikintjaku.'	Then someone'll say, 'Hey, maybe it's sweet enough to drink.'
Munu palunya nyaangka, tjanpingkalta kutanungka tjalatjura, kuuntjara arkara. 'Wiru!' Alatji.	And then, placing some greybeard grass in it, you try it, sucking it through the grass. 'Really good.'
Munu pulka mula tjuwita pulkaringkunyangka yungamalta, tjitji tjutalta, katira.	And when it gets really sweet, you take it and give it to the kids.

Kanytji

Likara nyangatja, kutjara maruarampa, ulupungkulampa.	This bark, you burn it, blackening it, and make it into a powder.
Palu pika palyalpailta nyangatja. Pika nyanga nyaa? Warutja, warutja nyanga.	It heals wounds. What sort? Ones from the fires — burns.
Nyanga palula marunika. Nyanga palula maru palurumpa — kamparalta patini. Pika ala nyanga.	You blacken the burn with the black powder — it stings and closes the sore up.
Pika alangka maru nyanga palula kuntjirampa.	You rub the black stuff into the open sore.
Ka kampanyilta, kamparampa patini.	And it stings and closes the sore up.

Shrubs

aratja — native fuchsia

Eremophila freelingii

A medicinal **ngangkaṟi** plant: the dried, resinous leaves are mashed up and mixed with water, and the resulting liquid is rubbed over the body, especially the head. Also used medicinally in the smoke treatment (fumigation) **puyutjunanyi** of people with body pain and bad coughs.

A small shrub 1–2 m high. Slightly hairy, resinous, narrow linear leaves grouped at the ends of branchlets, and scented when crushed; flowers are purplish, tubular with five terminal lobes; fruits green and globular with a single large seed.

Common around rocky sites, usually on rocky slopes, in small stands.

Murika

Mamungku patjaṉi, wankangku. Ka aratjangka puyutjunkupai.	If some live **mamu** are biting someone, you can give a smoke treatment with native fuchsia.
Ka patjaṟa patjaṟa — kutjuli ilupai.	It bites and bites them, and they all die.
Ka wati ngangkaṟingku uralpai. Miri tjuṯalta waṉira, wiyalpai.	Then the healer gets rid of them. Throws out all the dead bodies.

Palunya patjara nguwanpa iluntankupai palunya. Aratjangku iluntankupai kutjuli, wanka tjuta, mamu wanka tjuta	It sort of bites them and kills them. The native fuchsia kills off all the live ***mamu***.

Pompey Everard

Aratjangka puyutjunkupai.	You give a smoke treatment with native fuchsia.
Pika wai kuwaritja — pika una kutu, pika kutu ngaranytja ngaranytja.	Not for recent sickness by any means — for serious illness, pain that's been there for some time.
Puyutjura manarankupai, wiyalpai.	You give a smoke treatment to make it go numb, to kill the pain.
Kuntjultjara, kuntjulpungkula(3) pika pulkaringkunyangka, ngangkari wiyangku.	When you've got a bad cough, when you've been coughing and coughing; when you're really sick, and without a healer.

Kanytji

Aratja nyanga atura ulupungkula, kalpi nyanga, kapingka tjutira, kapingka wali-walirampa runyarampa, kuntjiralta anangu walytjangku kuntjira.	You pound the leaves of the native fuchsia almost to powder, mixing water with it, to mash it up, then rub it into the body.
Pikatjarangku kuntjira, kutjuli kutungku. Katangka tjunanyi yurulta katangka tjunanyi, munu ngarira ka palyaringanyilta.	A sick person rubs it in all over. You put the liquid all over your head, then while you're sleeping, you'll get better.
Ka ngaaly pulka, nyaa, Vicks puriny.	It's got a strong scent, like Vicks Vapour Rub.
Ka punu nyanga paluru kutulta, ngaalypa pulka, miritjina wiyangka, kuntjini nyanga palula, palyaringanyilta...	When we had no medicine we used to rub in this very plant, with the strong scent, and get better.

ilpara — silver-leaved water bush

Grevillea nematophylla

The lateral roots ***iwiri*** just under the ground are a source of emergency water ***kapi***, obtained by digging the roots up, breaking them into pieces and allowing them to drain into a bowl. Roots also have edible grubs ***maku*** of the ***katati*** type.

A tall shrub 3–5 m, sometimes growing to a large tree 8 m tall. Leaves narrowly cylindrical to about 20 cm long; flowers cream, in cylindrical heads to 7 cm long, toward the end of upper branchlets and very strongly honey-scented; fruit is an ovoid, lacquered-brown capsule, splitting longitudinally to release two winged seeds.

Occurs on sandplains, often associated with mulga. When ***ilpara*** is in full flower, Christmas is near.

Murika

Kulpara(3) kapitjiratja piltiringkupai.	Sometimes, on a long journey, we used to dry out for want of water.
Munu nyarkulta tjawara nyarakutulta ngarira(2) pakalpai.	We'd have to scoop aside the topsoil and lie against the cool earth for a while, before setting off again.
Munu yankula nyakupai, 'Talala yanu!'	And as you were going along, you'd notice something. 'There's a crack in the ground from a root!'

Ilpara paluru talala, talala ilpara. Paluru talala yananyi.	A crack in the ground, going off from a water bush.
Ka nyakula tjawani, munu katawakani.	And once we spotted that, we'd dig, and break the root by piercing it.
Tjawara wanara(3), katawakani, munu katawakani.	We'd dig along it and pierce one end, and then pierce the other end.
Munu kultu katakatakatantara wirangka, mimpungkananyi.	Then we'd break the root into separate pieces and put them in a ***wira*** dish, or rather a ***mimpu*** bowl.
Mimpungka mimpungka kanyini, ka mimpungka pulaparinyi, kapi paluru, ilpara paluru.	We'd leave them there, and water would fill up the bowl — water from the roots.
Kapi pulaparinyi, ka tjikira(2) kulyara(6) palunya.	The water would collect, and we'd drink it, and sprinkle ourselves with it to cool down.
Piruku iwiri kutjupa tjawalpai, munu pulapalta tjunkula, antjaki ngaripai.	Then dig up another root. And after leaving it in a bowl to build up, we'd camp overnight.
Tjikira(2) pakara kulpapai.	Then after drinking some more, we'd set out and return home.
Palu kapi pulka kutu — kapi anangungku tjikilpai, tjitjingku tjikilpai...	You see, it's an important source of water — water that Aboriginal people drink, that children drink.

Nganana yankula(2) nyawa. 'Ngangka! Wala pulkangku. Nyangatja maku. Ilpara nganmanyitja ngaranyi.'	We'd see one while we were on the move. 'Heck! Quick. Here's some grubs! There's an old water bush here!'
'Wanyula yuwala!' Munu ma-yuwara 'Ngangakaku! Ma-ma-ngarinyi!'	'Let's push it over!' And we'd push it over. 'Heck! There's a lot!'
Kala urara ngalkula(5) inalpailta. Inara wanara ma-inara wanara.	And we'd get them out, one after another, and eat them. We'd pull them out, all along the root.
'Ngura nyangatjampal tjutampa ngarinyi!' Punu kutjupa. 'Punu kutjupala nyangama!'	'Gee! There's so many here!' Another tree. 'Let's look at another tree now!'
Mai inara(2) katiralta warungka paura ngalkupai.	After pulling that food out of the roots, we'd take it, bake it and eat it.
Maku nyaratja, maku pulka kutu.	It's a source of edible grubs, an important source of edible grubs.
Ma-ngarira wanalpai, iwiri warara kutu. Ka nganana maku ngalkupai.	They're inside all along the roots. We eat those grubs.

ilykuwara

witchetty bush

Acacia kempeana

Witchetty grubs ***maku ilykuwara, maku lunki*** are a favourite and nutritious food source found in the lateral roots ***iwiri***. They can be eaten raw or lightly cooked. The roots are also used for making spears when the superior spear bush ***urtan(pa)*** is not available.

A multi-stemmed, grey–green shrub to 2–4 m high, or a single-trunked tree to 5 m high. Leaves smooth, elongated, rounded at the tip; flowers in yellow, cylindrical spikes; pods flat; seeds small, black, and ovoid.

A common species of open woodland and shrubland on sandplains and sandhills, sometimes forming small groves.

Murika

Nyangatja tjukutjuku ngaṟala(2), paluṟu patjaṉi.

After being small for a while the grub eats out the root.

Patjaṟa, ngarira(2) ma-puḷkaringkula puḷkaringanyi.

It eats the root, lying inside it, getting bigger and bigger.

Munu paluṟu puḷkaringkulampa, lipiringkulampa, karuṟany, puṉu karuṟanypala watjaṉi. Paluṟu, piṯi nyanga ngura paluṟunku, lipiringkula.

And when it gets quite large, quite wide, we say **karuṟany**. For its tunnel in the root, its home, once it swells.

Munu paluṟu ngarira(2) impilta aḻaṉi. Impi aḻaṟa, paluṟu ngarira(2) pakaṉi, mirinarira.

After lying there for a while it makes an opening in its tunnel. Then it comes out, as an immature grub.

Piltaly paluṟu piltalytjunanyi. Paluṟu nyaḻpitjararingkula yananyi.

It sheds its case. Gets wings and goes off.

Kala nganaṉa impi wakaṟa aḻaṟa nyakula, iṉaṟa(6) nganaṉa katipai.

As for us, after we pierce open their silken tunnels and find them, we pull out a whole lot of grubs and take them with us.

Ka kutjupa puṯu wakaṟa, puṉu nyaratja kantilya, wiya, maku ngarinytja wiya.

If you can't pierce open a root — if it's too hard, there'll be nothing, no grubs.

Palu piltalypa nyakula, karuṟanypa nyakula, wakaṟa parpaṉi.

But if you see the cast-off cocoon cases, if you see swellings in the roots, if as you tap the ground to test you hear something...

'Munta, nyangatja tukulmanayi!' Ka palunyalta iṉaṉi, tjawaṟa.

'Oh, it's giving out a hollow thud here' — you'll be able to get grubs out, once you've dug.

maku tjawani: *digging for witchetty grubs*

irmangka-irmangka — native fuchsia

Eremophila alternifolia

A much-valued medicinal plant: the leaves are chopped and mashed to a paste, and used as a 'rubbing medicine' for the head, or put into grass and tied onto the head as a poultice.

A shrub to 3 m high. Leaves aromatic when crushed, linear, almost cylindrical in shape; flowers typical native fuchsia type, pinkish and variously spotted inside and out; fruit ovoid-conical, smooth.

Occasionally found on the rocky slopes of hills and ranges, and on rocky outcrops. Can be confused with *Eremophila latrobei*, another native fuchsia common in mulga country.

'Pika, yaaltjiku nyangatjar? Ngangkari nyanga wiyangkampa?' Palu 'Punu palakutunti ngaranyi?'

'Oh, what is there for this illness? Is there no healing agent around?' 'Wouldn't there be a herb around?'

'Irmangka-irmangka nyangatja ngaranyi. Nyanga aputja — ila. Punu nyanga ilatja kutu ngalya-ngaranyi.'

'There's some native fuchsia, on the rocks nearby. There's some really close.'

'Walangkuya katantara! Nyanga paluru wankalpai, mamutja.'

'Go get some quickly. That stuff cures those stricken with ***mamu***.'

Ka anangu kutjupangku yankula katantankupai. Katantara katira.

So someone goes and gets it. Breaks some off and brings it back.

'Uwa, alatji palyalta pala palurumpa.'

'Yes, that's the way, that's the right stuff!'

Atura, runyura mulatu, kaputura, kapingka kaputura katangka alatjingara tjarutjura, tjarutjura.

Then you chop it up, pulp it up well, make it into a poultice with water, and rub it like this on the head.

Nyitira(2) kata nyitira(7), nyitiralta katangkalta nyanga kutulta tjunkupai.

You rub it into the head, really rub it in around here.

Kaputulta, irmangka-irmangka kaputu, tjunkulampa karpiralta wanti.

You tie the poultice onto the head and leave it.

Ka paluru ngariralta pikatjara palyariwa.

Then the sick person sleeps, and gets better.

Pakaralta nyinamar, 'Wai, pikampa palyaringu, ipily nguwan ngaranyi.'

He gets up, and sits up. 'Hey, the pain has gone! I'm almost healthy again.'

'Ngura palyaringu kutuna! Punu nyanga paluru wankalpai kutu. Palya!'

'I'm just — fully recovered. This plant really does cure you! Great!'

Ka anangu kutjupangku nyakula watjalpai. 'Kukampa nyarangku wati-nyinara ngalkunir. Palyaringju manti.'

Other people look and say, 'Gee, that person's sitting up and eating meat. Looks like he's better.'

Munu wankani, punu nyangangku, irmangka-irmangkangku 'Muntawa palya.' Paluru palyaringkupailta.

It cures, this native fuchsia. 'Well, I'm better.' He recovers then.

Munu 'Paluru kaputu panyalta arara wanira nyinamar!'

'He's taken the poultice off and thrown it away. He's sitting up!'

Tjintukutu pakara. Paluru kaputu winki

This is after getting up the next

ngarira tjinturingu, panya nyangakutu tjunkula katangka.

morning. He'd spent the night with the poultice on, that he put on his head.

Paluru pakaralta para-ngarala ngura yankupai kutu. Para-ngaramalta. Alatjika.

He gets up, walks around and just goes off. He walks around normally. That's enough.

Pompey Everard

Irmangka-irmangka, punu paluru, kalpi pala atulpailta.

You pound up the fuchsia leaves.

Atura kapingka kulya-kulyara, yurura mulatu, putja tjanpingka tjunkula ka katangka tjunkula.

Chop them up, moisten them with water, and when it's nice and wet, place it in a spinifex grass and put it on the head.

Ka karpira walura mulatu, ngaripailta.

You tie it on really securely before going to sleep.

Munu raapamilalpai, kuntjilpai. Kuntjira, ngarira, tjinturingu munu putja panya waniwiyangku wantima.

You also rub it in. Rub it in and sleep the night. You don't throw that grass away, but leave it on.

Munu paluru ngarira tjinturingkula kulilpai. 'Uu, ngura pika panya wiya kutu ngaranyi.

After sleeping the night, you'd think: 'Hmm, well, there's no pain at all.

'Ai, palyamantir! Punu nyangatja miritjinampa.'

'Hey, looks like I'm better! This plant really is medicine.'

karingana

mint bush

Prostanthera striatiflora

A medicinal plant ***ngangkari***: pounded whole and applied to the head as a poultice ***kaputu***, or rubbed onto the chest. The poultice may also be wrung out, and the resulting liquid sprinkled over the body.

A low shrub to 2 m tall. Leaves are smooth, opposite, to 2.5 cm long, and fragrant when crushed; flowers tubular, lobed and white with yellow spots in the throat, and purple lines inside the tube.

Usually found in sheltered places on rocky ranges, hills or outcrops.

Pompey Everard

Karingana atura, yurura, kaputura katangka tjunkupai, munu nyitilpai, pilpirta.

You mash up the mint bush, make it into a poultice and put it on the head, and rub it into the chest.

Putjangka kaputu tjunkula kuwitilkini. Kapi yuru tjutini anangu pikatjarangka. Kulya-kulyani.

You put the poultice into some grass and wring it out, so as to drip the liquid on the sick person's body.

Ka paluru palyaringu.

And they recover.

kupaṯa

wild plum, bush plum

arnguḻi

Santalum lanceolatum

An important plant food ***mai puḻka***. The skin and flesh of the fresh fruit is eaten raw. The fruit is also mashed with water and drunk. The dried fruit may be reconstituted with water and eaten. The wood is used for carving small animals.

A tall shrub or small tree to 4 m with pendulous branches. Leaves fleshy, bluish, to 5 cm long and 3.5 cm broad, tapering at both ends; flowers in clusters at the end of branchlets, each made up of 4 yellow–green fused petals; fruits shaped rather like a small gumnut, purple–black, with a little flesh covering a single large seed; a root parasite.

Often in small groups, along watercourses, or in places such as the base of rocky outcrops where water runs off. Also on sandhills and sandplains. Fruiting after favourable rains.

Milatjari

Kupaṯa nyangatja mai mulapa. Mai panya mangaṯa purinypa, mangaṯara.

The wild plum — it's an important food. It's similar to the quandong. There are the two of them —

Alatjiṯu pula, paluṟu pula, arnguḻi. Munta, kupuṯa — arnguli, ini kutjara.

the quandong and ***arnguḻi***, sorry ***kupaṯa*** — there are two names for wild plum.

Uwa, nyara palunya uraṟa kuḻu-kuḻu ngalkula, ngalkula kuḻu-kuḻungku ngalkula, tjatjili-tjatjilingku.

Yes, you gather them up, eating some as well as you do.

Palunya uralpai, mimpungka, urara(3), tjunkula(4) ka pulaparipai.

You gather them up in a ***mimpu*** bowl, putting in more and more till the ***mimpu*** gets full.

Ka ngurakutulta tjalira katipai, ngalkula kulu-kulungku.

And then you carry it on your head to camp, eating some on the way.

Pika mulya wirunya alatjitu! Nyaratjatu kaika purinypa. Mai paluru, ananguku mai.

It's really good dry — like cake. It's Aboriginal food.

Katiralta ngura kapi tjutiralta tjunyilpai.

You just take it and get some water, and soak it.

Munu ngura wiltjangka tjunyira(4). 'Ngangkaku' — ngalkula kulu-kulungku.

Squishing it around in water, back in the shade. 'Heck yes' — eating some as well.

Wiya, mai panya wirunya alatjitu. Kungka tjutaku mai.

No worries; it's really good food — women's food.

Ngayulu mai nintitu palumpa. Mai wiru nyaratja.

I really know that food. It's excellent.

Palawa wiyangkala nyara palunya ngalkupai.

We used to eat it before we had flour.

Tjunyira ngalkula(2) ma-ma-ngarima, munu pakara tjali-tjalilpai, mimpu.

You soak it, eat some and lie around the place, and then get up and load the ***mimpu***s on your heads.

'Arnguli panya palumpatu puta wanyu yara, piyuku!'

'What do you say we just go back to those same wild plums again!'

Ka wati tjutangku kuka kanyala wanalpai. Ka kungka tjutangku arnguli uralpai.

The men used to chase euros. And the women would get wild plums.

Urara(4), wiltjangka katira, tjunyira ma-ma-nyinama.

They'd get lots, bring it all back to the shade, soak it and be sitting around.

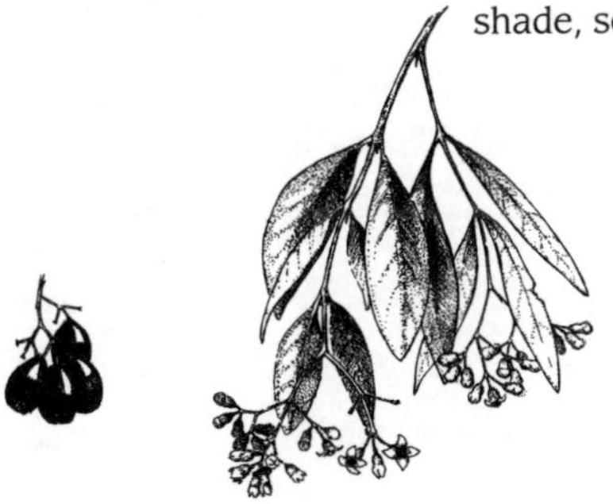

Ka ngura tjitjingku 'Ngunytju, ngunytju! Ngalya-yuwa<u>n</u>i, arngu<u>l</u>i.'

And the kids would say, 'Mummy, mummy! Give me some wild plum.'

'Wiya, palatja ngalkunma, nyuntu panya ngalku<u>n</u>ingi, pu<u>n</u>ungka panya.'

'No,' she'd say. 'You eat that over there, what you've been eating before, off the tree.'

Ngalkula ngalkulalta ngura wiltjangkalta nyinara wa<u>n</u>imar.

They'd eat and eat, and be just sitting around in the shade.

Ka wati tju<u>t</u>angku kukalta katira, kanya<u>l</u>alta wa<u>n</u>i-wa<u>n</u>ipai.

Then the men would bring some euros and throw them down.

Wati tju<u>t</u>angku ngapartjilta ma<u>l</u>ulta pau<u>r</u>a(2) ngalkula(2) 'Ngangakar!' Mungartjirira, wangkara wa<u>n</u>ima.

The men in turn would roast kangaroo and have a good feed as it got late, sitting around talking.

Tjuka, palawa, tilipi wiya. Palunya kutju, kuka kanya<u>l</u>a arngu<u>l</u>i ngalkupai, ngana<u>n</u>a. I<u>l</u>i ku<u>l</u>u-ku<u>l</u>u, mai i<u>l</u>i ku<u>l</u>u-ku<u>l</u>u.

There was no sugar, no flour, no tea. Only that. We used to eat euros, and the wild plum. And the wild fig as well.

Uwa, nyanga paluny<u>an</u>a wangkanyi, kutju.

Yes, that's all I've got to say.

tjunta<u>l</u>a

colony wattle

Acacia murrayana

The mature, hard, black seeds are separated from the pods, moistened with water to make them soft and then ground to a paste ***latja*** which is eaten. Alternatively, the seeds are parched in a wooden dish ***ka<u>n</u>il(pa)*** with hot sand, ashes and coals, separated by yandying, and then ground. Edible grubs ***maku*** are found in the roots.

A tall shrub or small tree to 5 m high. The trunk and branches often have a greyish–white colour contrasting with the grey–green foliage; leaves are narrow-linear with a faint but discernible midrib; flowers in yellow, globular heads, pods dry to a shining tan colour; seeds are small, ovoid, with a characteristic depression in both sides.

A common species of some areas occurring on sandhills, sandplains, and along watercourses and floodplains. Known as colony wattle in English because it may spread from underground roots and produce small groups or colonies. Fruiting in spring and summer.

Sam Pumani

Nyulkuṟa, nyulkuṟa mai palunya kaṉinilta.	After rubbing it to open the pods and free the seeds, you yandy it.
Kaṉiṟa, palunya wirangka tjunkupai.	After yandying it, you put it in a ***wira*** dish.
Wirankga tjunkula kulya-kulyaṉi. Mai palunya kulya-kulyaṉi, kumpulingka, tjulaṉi.	You put it in a ***wira*** dish and moisten it. You sprinkle the seed with water and soften it.
Tjulaṟa, tjiwa tjunanyi. Munu tjiwa tjunkula mai palunya tjunanyi tjiwangka.	After softening it, you get a flat lower grindstone. And having put the grindstone in place, you put the seed on the stone.
Munu palulanguṟu tjungarilta mantjini. Tjungaṟi mantjiṟa, rungkaṉilta. Rungkaṟa latjapunganyi.	Next you pick up a millstone. You take a millstone, and grind it to a paste.
Latjapungkula ngalkuṉi, ngalkula	Make it into a paste, and eat it — eat

talturingkula, talturingkulampa... talturingu ngalkula(2).

till you're full.

Tjiwa tjunkula wantira, ngurangka nyinanyi, mungartjirira, waru tjunkunytjikitja.

You eat for a while, put the millstone aside, and sit in camp, as it's getting late, and you want to make a fire.

Yuutjura, yuutjura ngarinyi, mai palunya ngalkula(2) taltu.

You make a windbreak and sleep, after eating that food — full.

ultukun(pa) — honeysuckle grevillea

Grevillea juncifolia

The orange blossoms are a source of honey ***tjuratja***, ***wama***, which is sucked directly from the flowers, or flowers are soaked in water to make a sweet drink.

A shrub to 6 m tall, usually with two to several main upright stems, dividing just above the ground. Needle-like leaves simple or divided, to 27 cm long; flower spikes orange; clustered pods slightly hairy and rounded, with a small spike; each pod has two seeds, each with papery wings.

On sandhills and sandplains throughout the region, often forming tall open shrubland, or scattered in other formations. Flowering in winter and spring if conditions are favourable.

Ka ul̲tukunpa kat̲antar̲a kapingka tjunkupai.

You break off some honeysuckle grevillea and put it in water.

Munu tjal̲atjur̲a(5) — tjunyintja wiyangku — pur̲inytju mar̲ungku nyurmintananyi, alatjinanyi.

After immersing a good deal of it — without squishing it around at all — you squeeze it gently with the hands, like this.

Munu nyanganyilta. 'Nyangatja mar̲uringu.'

And watch it. 'It's gone dark.'

Munu kulini 'Tjingur̲u tjikintjakulta wamalta, wir̲ulta. Nganan̲a tjikintjaku ngar̲anyi.'

And think: 'Maybe it's got sweet enough to drink. It's ready to drink.'

Munu tjanpi piyuku tjunkula, katu tjunkula, raputji tjikilpaingka, kilinananyi.

And you put some grass in on top as a strainer, in case you should drink the rubbish.

Munu palunytjanungku, panya palur̲u pul̲ka kutjupa, palur̲u pur̲inypa tjikilpait̲u.

And after that, you drink the other big one just the same.

Tjunyir̲a(6), tjal̲atjur̲a(4) tjunkupai, munu wati tjut̲a 'Panya, mapalkungku! Rawangku tjunkunytja wiya!'

Squish it around, immerse it and put it aside, and the men would say, 'Come on! Quick! Don't let it sit for too long!'

Palur̲u ngarira panya nyaaringkupaingka — un̲aringkupaingka, raputjiringkupaingka. Uwa, tjuwita wiyaringkupaingka.

Lest it stand and go off. In case it spoils. Yes, lest the sweetness disappear.

waputi — desert thryptomene

pukara — *Aluta maisonneuvei*

A traditional source of nectar ***tjuratja***, ***wama***. Early in the morning before sunrise, the nectar and dew is beaten from the flowers with wooden ***wira*** dishes, and collected in ***mimpu*** bowls to drink.

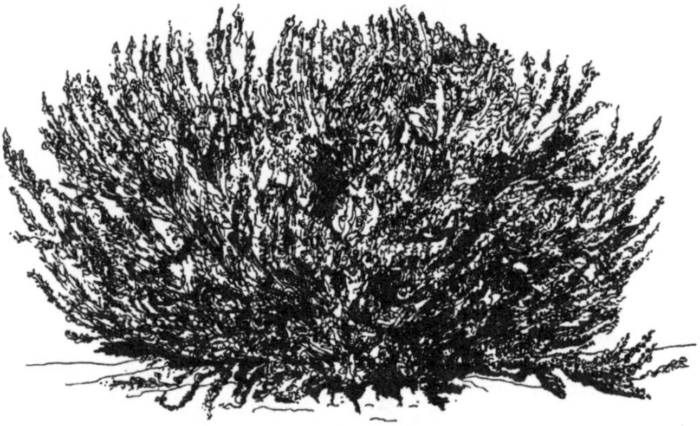

A low, much-branched, rounded, compact shrub to 1.5 m high. Minute leaves crowded along the end of branchlets, aromatic when crushed; bark peeling in thin red strips; flowers small, with white or pinkish petals; fruits are small woody capsules.

A dominant shrub of sandhills in some areas. Flowers in winter and spring.

Milatjari

Ngayulu tjitjingku nyakupai, ngayuku ngunytju. 'Yankulala panyangka panyalta puta ma-tjura!'

As a child I remember seeing my mother say, 'What do you say we take our dishes and things to that same place as before!'

Nyinngangka wama, paluru nyinngangka ngarapai, tjuka paluru, tjinkintjaku. Kala ngura katiralta ma-tjunkuku, talingka parari-pararilta.

The nectar is there in winter, ready to drink. So we'd take our stuff off right into the sandhills.

Munu nyakupailta 'Wampanti wampanti. Tali kutjupa, tali kutjupa. Wampanti, ngankar!'

Kala ngura pika parari ngaripai, tali kutjupangka parari-parari.

Ngariralta panya. Piluka kutju mungawinki katipai, paluṟu puṟinypa pakalpai, aṉangu ngapartji.

Munu pakaṟalta mimpulta uralpai, wira tukulkulta, mimpu wira tukulkulta.

Ka kutjungku waraṟalta pungkupai, pungkupai, arkalpai pungkula, mungangka, aḻipala!

Ma-pungkula 'Ngangkapa, parkanguṟu ngalya-pupakatipai,' wapuṯi palunya, tjuka wapuṯi.

Alatjiṯu pungkula(4) ma-kaḻaḻarira 'Pau! Pau! Pau!'

Palunya wangkapai, wangkapai, mirara wangkapai.

Ka kulilpai, 'Ngangkaku, nganampa kunyu wapuṯi puḻka pungu.'

Munu alatjiṯulta wira tukulkutjara tukulkutjara.

Munu ngura pungkula(2) alatjingaṟa pungama, parka, pungkupai.

'Pau, Pau.' Pararitja, pararitja, pararitja. Tali kutjupitja, tali kutjupitja.

Mimpungka pungkula tjaṟuwaṉira(2), pitingka panya, pungkula tjaṟuwaṉira(2) yuṟu, kapi puṟinypa.

Pungkula parkangka pungkula tjaṟuwaṉira(3) ka mimpungka panya,

And we'd see the desert thryptomene, 'I don't know what to say. It's all over the sandhills! I'm lost for words!'

And we'd camp out there, a really long way off, between some sandhills.

We'd sleep. And you know what it's like moving cattle in the early morning — we'd get up then.

We'd get up and fetch our **mimpu** bowls, and cup **wira**s.

Someone would test the bushes first, while it was still dark — really early!

They'd be off beating at the bushes, singing out, 'Heck, the dew's dripping off the leaves,' the sweet thryptomene dew.

We'd beat the bushes till the sun got hot, singing out 'Pau! Pau! Pau!'

Shouting it out loudly, so the others would know.

They'd hear and think, 'By gee, it sounds like they've beaten out plenty of sweet dew for us.'

You do it just like this, with a cup **wira**.

They'd just beat at the bushes like this, beat the leaves.

Shouting out 'Pau! Pau!' Getting further and further away, going from sandhill to sandhill.

They beat the dew out, and tip it in **mimpu** bowls, the liquid, like water.

They'd beat if off the leaves, and tip it into **mimpu**s and those **mimpu**s, those

pitingka panya, pulapari!	***piti**s*, would fill up with it!
Ka wirangka ngapartjilta pungkula tjikinma.	They'd also have a drink straight out of the ***wira***, as they were doing it.
Pungkula tjikilpai uṟu, pungkula tjikilpai.	They could drink some as they were getting it.
Munu katira waṟungkalta tjura, wampanti wampanti.	Then take it back to the fire — so much of it!
Munu palunya, tjuṯa palunya pungkula(2) puṉu palunya pungkula, pungkula(2) kulpa.	After beating the sweet dew, out of plenty of thryptomene bushes, they'd go back.
Katiralta tjunkupai mimpu panya palulampa, piti panya palulampa.	They'd take it back in the ***mimpu**s*, the ***piti**s*.
Katira tjunkula waṟulta kutjalpai.	They'd take it back and light some fires.
Munu pika nyaangka, alṯa nyanga puṟinytja, nyanga puṟinytja, alṯa nyanga puṟinytja alatji ngaṟaku, alatjinkupai, miṉa.	And then, with a coal like this — they'd do this to their arms, scrape the sticky dew off.
Palunya pungkula tjikiṟa(2), alatjinkupai, maṟulpai.	After all that beating it off, and after having a drink, they'd do this to their arms, blacken them by scraping with a coal.
Ka tjikiṟa tjikiṟalta ngura waṟulta nganytjiṟa(2), waṟu wiyatjangku.	Then after a good long drink they'd warm up by the fire, the ones who'd been without fire in the cold.
Pungkula tjikiṟa katiralta tjunkula nyinara, kalaḻarira ngura kunkunpalta.	After all that beating and drinking, they'd stay in camp as it got hot, and just sleep.
Panya paluṟu tjanampa, nguurmara waṉima. Kukaku ankunytja wiya. Panya palunya, tjuka panya palunya pungkula tjikiṟa(2).	They'd lie around snoring. They wouldn't go for game, not after beating out and drinking all that sweet stuff.
Munu kuṉtjulpungkunytja wiya. Wanka panya alatjiṯu nganaṉa nyinapai.	And we didn't use to have coughs and colds then. We were really healthy.
Nyanga palunya kutjuṉa wangkanyi	That's all I've got to say.

wat̲arka — umbrella bush

Acacia ligulata

A food source: the hard seeds are soaked in water, or parched in a wooden dish ***kan̲il(pa)*** with hot sand, ashes and coals, separated by yandying, and then moistened. They are then ground to a paste ***latja***, which is eaten. It is said that the food makes your hair fall out. It was mainly used in drought times ***ailurungka***. Edible grubs ***maku*** are found in the lateral roots ***iwiri***.

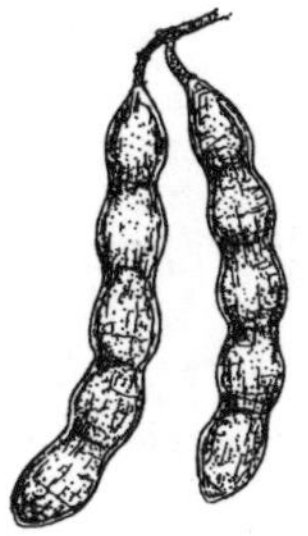

A tall multi-stemmed shrub to 5 m high. Leaves a little fleshy, variable in shape but generally narrow-linear, rounded at the tip, with a distinct midrib; flowers yellow, globular; pods thick-walled, slightly constricted between the seeds; seeds black, oval, centrally depressed, with a prominent red or yellow appendage.

A shrub of sandplains, sandhills, and sometimes sandy watercourses. Fruiting spring and summer.

Mai mulapa kunyu nyaratja.

The umbrella bush is said to be a true food.

Palunya kunyu muurpungkula tjunkupai.

They say you crush it up and put it aside.

Munu kunyu waru ipangka paura, wiruralta, kanirampa tjiwangkalta rungkara ngalkupai, tjungaringkalta.

And after parching it in the hot ashes, yandying it really well and then grinding it on a grindstone, you eat it, after grinding it with a millstone.

Mai mulapa kunyu watarka, munu maku ngarapai iwiringka.

It's said it's a genuine food, and there are edible grubs in the roots.

Ka maimpa ngarapai, nyaampa? Mai panya punu nyaangka? Parkangkampa.

Where on the bush is the food part found? Among the leaves.

Ka kunyu mai mula. Miri tjutangku kunyu ngalkupai mai palunya.

It's said to be a true food. The people who've passed away used to eat it.

Munu kunyu kata pampulpai, maitjanungku. Ka kunyu uru punkalpai, munu kunyu ngula pakalpai.

And, it's said, they'd touch their heads after eating it. And their hair would fall out, and grow again later.

Munu kutjuli punkara wiyaringkunytja nyinara(2) kunyu uru ngula pakalpai.

They say once the whole lot had fallen out, after a while it would grow back, in time.

Punu mulapa nyaratja. Maitu kunyu paluru. Ka makumpa iwiringka ngarapai.

It's an important plant. It's said to be a plant food, too. And there's edible grubs in the roots.

Subshrubs

iṯuny(pa) — **bush tomato, western nightshade**

kumpul(pa) — *Solanum coactiliferum*

A food plant: when ripe and yellow, the small berries may be eaten raw, once the bitter juice and seeds have been expelled by squeezing or piercing with a stick. Alternatively, the fruits are chopped open, and once the bitter juice and seeds have been cleaned out, they are baked and eaten.

A small erect, hairy subshrub to 30 cm high. Often with prickles which are recurved and usually present only on the stems; leaves oblong, often folded, to 5 cm long; flowers purple with central yellow stamens; fruits yellow when ripe, to 1.5 cm in diameter.

A common plant of sandplains and sandhills, often found in groups. Fruiting after favourable conditions.

Murika

Talingkala yankupai, munu nyakupai taliwanungku.	We'd be travelling in the sandhill country. And across a sandhill, we'd see some.
'Ngangarku! Palatja pintalynyinanyi, pala!'	'Oh heck, it's all over the place just there!'
Kala pintalynyinanytjala uraṟa(6) iṯuny palunya. Waṉaṟa uralpai.	And so we'd gather and gather it, for a long while, that bush tomato. We'd gather it as we went along.

Munu tjar̲uwan̲ira(3) katira, nyakupai. 'Nyangatja kampur̲ara ngar̲anyi.'

After dropping lots and lots into our dishes, as we were taking it off with us, we'd see something else. 'There's some desert raisins!'

Kala kampur̲ara piyuku uralpai. Urar̲a, urar̲ala kampur̲ara it̲uny katipai.

Then we'd start gathering desert raisins. After getting some, we'd take both the desert raisins and bush tomatoes off with us.

Munula katira tjunkula, it̲uny paur̲a wantipai. Munula wakar̲a ngalkula(2).

After bringing them back to camp, we'd put the bush tomatoes to bake. Then we'd pierce them open and eat them.

Mai nganampa nyangatja. Nganan̲a mamutjara wiyangku ngalkupai. Kurunpa wankangkula ngalkupai, pikatjara urkalypa wiyangku...

It's our kind of food. We used to eat it free of ***mamu*** spirit-monsters. Eat it with a healthy spirit, not sick, free of colds.

Uwa, paur̲ala ngalkupai, palunya, wakar̲a. Pun̲ungka wakar̲a, ngalkupai.

Yes, we'd bake and eat it. Pierce it with a stick and eat it.

Purkar̲ala tjunkupai. Wirangkala wakar̲a tjunkupai.

We used to put it to cool. We'd pierce it, then put it to cool in a ***wira*** dish.

Nyara palunya wakar̲a tjunkula(2) ngalkupai. Mai nganampa nyaratja, ngal̲tutjara.

We'd pierce the fruits, put them to cool, and eat them. That dear old stuff is our food.

Palur̲u tjana atur̲a tjulkulpai. Apungka atur̲a tjulkulpai, munu paulpai. Munu kilina war̲u ngalkupai. Mai wir̲u nyaratja.

They'd chop it open and squeeze the juice out, chop it with a rock and squeeze it, then bake it. And eat it really clean. It's fine food.

(Nyaakuya putja tjunkupai?) Putja tjunkupai, munu kilinangka paulpai, nyaaringkupaingka, pil̲tiringkupaingka. Mantaringkupai. Munu palur̲u wir̲ur̲alta ngalkupai...

(Why did they put grass in place, in the fire?) They'd put grass in place, and bake it in clean grass. Lest it dry out — and so it wouldn't get dirty. Then they'd eat it, relishing it.

Munu yankula nyakupai. 'Nangatja kutjupala kura arkar̲a wantima, tjuwita wiya.' Kura arkar̲a wantima.

Sometimes they'd go find some, and say, 'Let's leave this stuff, since it tastes bad, not sweet.' If it tasted bad, you'd leave it.

'Nyangatjampal tjuwitampa!' Tjuwita palatjampa uralpai. Mai wir̲u arkar̲ala

'Oh, this here's sweet!' We'd gather just the sweet stuff. After tasting it was

urara(3) kulpalyinkupai.

good, we'd gather plenty of it and take it back to camp.

Munula paura ngalkupai palunya, mai. Anangu tjuta kutungku paura ngalkupai.

Then we'd bake it and eat it. All Aboriginal people used to bake and eat it.

Wati miri tjutaku mai nyaratja. Mai wirunya.

That's the food of the men who've passed away. It's really good food.

kalpipila

Crotalaria eremaea
ssp. *strehlowii*

A common medicinal plant ***ngangkari***: it is pounded, mashed and rubbed into the head and chest for general illness. Alternatively the pounded plant is put into ***kutanu*** grass and tied on the head as a poultice.

A small, spindly, open subshrub to 1 m high. Leaflets smooth, oval-shaped, folded to 8 x 3.5 cm; yellow pea-like flowers on terminal spikes; pods to 3 x 0.7 cm, beaked at the tip, with up to eight seeds.

Mostly found on sandhills, sandplains and watercourses in mulga woodlands and mixed shrubland.

Murika

Nyanga paluru, kalpipila, nyanga atura tjunkupai, kamilu, katangka tjitjingka.

A grandmother pounds this up to apply to a sick child's head.

Raapamilalpai nyangatja, kututu kulu-kulu raapamilalpai, pika ngarala. Ka nyanga paluru palyalpai.

It's rubbed in here. Rubbed into the chest also, where there's pain. It makes you better.

Nyanga paluru ngangkari, ngangkari pulka nyangatja.

This plant is an important healing agent.

Takata wiyala nyanga paluru ngangkaringka wankaringkupai.

Before we had doctors, we would recover with the aid of this healing agent.

Pompey Everard

Irmangka-irmangka wanma ngaranytjala, nyanga palunya ngaranytjala, atuni.

If the mint bush is too far away, but this plant's around, you chop it up.

Munu tjanpingka katu tjunanyi, kutanungka, katangka, putjangka, munu karpira, wantinyi.

And put it, in some grass, in some greybeard grass, on the head, tie it on and leave it.

Ka ngarira tjinturingkula pakara nyinara, palya kutu ngalkunyangka kulinma:

And after she's slept the night, sits up, and is eating well, you might think:

'Ai, tjitji nyangatja manti palyaringumpa. Mula manti. Wirumpa, miritjina wiya.'

'Well, looks like the child is better. Looks like it's true! That's great, and without medicine.'

Herbs

mingkul̲(pa) — **wild tobacco, pituri**

pulyan̲t̲u
ukir̲i

Nicotiana excelsior

Highly prized, and much sought after for its narcotic properties. Whole plants are harvested and the dried leaves are crushed, and mixed with the ash of the bark, twigs, or leaves of certain species, e.g. the bark of river red gum ***apar̲a***. This mixture is then put in the mouth, rolled into a quid ***kaput̲u*** and sucked on.

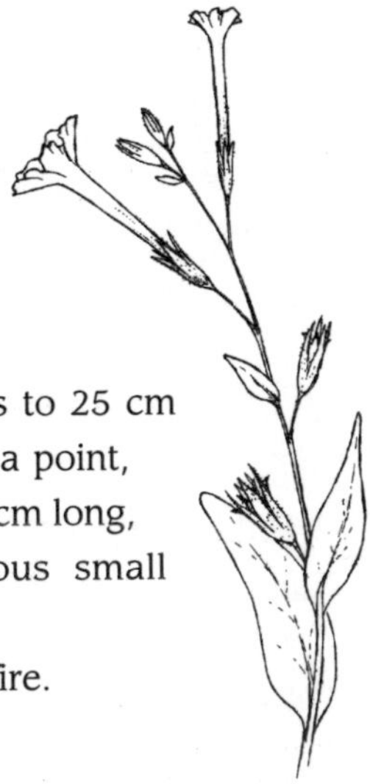

A leafy herb to 2 m high. Large leaves to 25 cm long, slightly fleshy, smooth, shiny, broad, tapering to a point, and stem-clasping; flowers white, tubular, narrow, to 8 cm long, fluted at the opening; capsule containing numerous small brown–black seeds.

Occurs infrequently in rocky soils, commonly after fire.

Sam Pumani

Pulyan̲t̲u ngal̲t̲utjara, kal̲pi wiyangka 'Kal̲pi wiyariwa.' Ka miri tjut̲angku tarkalta tjunama.	Suppose they were poorly off for tobacco, without any leaf. 'We're out of leaf!' In this case the people who have passed away would turn to the stalks.
Munu tarka runkar̲a ngapunma,	And grind them and chew them dry,

nyulytjanma, patjanma. Patjalpai nyangatja.	suck on them and bite them.
'Wanyu wanti kana patjala. Tarkanatju patjantjakur, patja-patjantjaku. Ngayulu tjaa piltiringu.'	'Well, I'll just have to chew this. I can sort of chew the stalks! My mouth's gone dry.'
Ka 'Ngayulu nyangatja wiyangka tarka katirinanyi.'	'Since I don't have any leaf, I'll just have to carry stalks around with me.'
Tarka miri tjutangku katirinkupai kalpi wiyangka,	The people who have passed away would take stalks around with them, when they didn't have the leaf.
munu tarka palunya rungkara kaputunkupai. Ulupungkula rungkara kaputu katirinama.	They'd grind the stalks into a quid. After grinding them to powder they'd carry the quid around with them.
Ka tarka tjuta ngarima, wipiyangka. Wipiyangka ngaripai. Wipiyangka tjunkupai, emu panya.	The remaining stalks would be kept wrapped in emu feathers. They put them in emu feathers.
Ka rungkaralta purinymara ngalkula(2) ngalkukatira ngalkukatira ka apu ilaringkula pulyantulta ilariwa.	And after grinding and softening them, they used to chew them, stringing it out to make it last. And as the hills got close, fresh tobacco would get close.
Ka tarka panya palunya waninyi. Uwa, tarka waninyi.	Then they'd throw away the stalks. Yes, throw the stalks away.
Pulyantu panyalta katantananyi, munu palunytjanungku witani kalpi witani, warungka panyalta.	Break off some fresh tobacco, and after that, singe the leaves on a fire.
Witarampa 'Wanyu tjalpilpi nyulytjani', nyulytjanilta, pulyantu wiyatjarangku.	As they were doing that, one might say: 'Hold on, I'll just suck on a fresh leaf!' He'd suck it all right, the one without tobacco.
Uwa, kurulta pakala, pulyantu ngalkula.	Yes, then his eyes'd lift up, from chewing that tobacco.
'Uwar, kanatju pulyantu ngalkula! Ngayulunatju pulyantu wiya nyinanytja nyinanytja.'	'Oh yes! I'm chewing tobacco at last! I've been without it for ages!'
'Walawi kanatju tjaali ngarir!' Tjaali ngarinytja wangkama.	'Oh, quick, and I can sleep with it in my mouth!' They would talk about sleeping with it in the mouth.

Uwa, munu... 'Tjaali̲lta ngarima! Mungaringkula, mungaringkula tjaali̲lta ngari!'

'Oh, to sleep with the stuff!' As it gets dark, they'd be saying: 'Oh! I'm going to sleep with it!'

Tjaali̲lta ngarinyi, kaputu̲ tjaali̲lta, kunkunpalta ngarinyi. Tjaali̲, wiya ngarinytja wiya.

And he'd sleep with it, sleep holding it in the mouth. He doesn't sleep without it.

Munu tjin̲turingkula pakar̲a(2), pinangka tjunkula katirinama.

When day breaks, they would get up and tuck it behind the ear, and carry it around with them.

Munu tjaali̲tjara pukul̲pa anama, pukul̲arima. Tjaangkananyi munu tjaali̲ ananyi, kuka ngurilkatinyi.

And would travel around contented, with it held in the mouth. They'd feel satisfied. They put some in the mouth and travel around, looking out for game.

parkily(pa) — parakeelya

nyurngi — ***Calandrinia balonensis***

A food plant: the whole plant is baked in hot earth and ashes, and the leaves eaten. Water source: the succulent leaves are broken open and sucked.

A small, annual succulent. Ascending leafless flower stems rise from a basal cluster of leaves; leaves cylindrical, to 4 cm long; flowers usually pinkish-purple with five separate petals; fruit a small conical capsule with numerous small seeds.

Found in a wide range of habitats on a variety of soils. After rain, it often covers large areas on sandplains, commonly in association with mulga.

Mai wiru nyangatja parkily, nyanga paluru.	It's fine food, this parakeelya.
Ka mai palunya nyangatja ngalkuni, ala nyangatja.	You eat this part of it. This part here *[indicating leaves]*.
Paura, warungka paura, alatjingara(2) ngalkuni, wata wiyangka.	You bake it in the fire and eat it like this, without the stem.
Munu mina nyangatja wantinyi. Waninyi mina nyangatja, ngalkula(2).	You leave the branches alone as well. You discard the branches, as you eat the leaves.
Mai wiru kutu nyangatja, nyanga paluru. Mai miri tjutangku ngalkupai.	It's really good food. The old people who have passed away used to eat it.
'Mama, ngunytju. Parkilytja yuwa!'	'Father, mother. Give me some parakeelya!'
Watjantjala, nyanga alatjiya katantara yungkupai.	When the children spoke like that, they'd break some off to give, or they'd say:
'Wiya, nyangatja mai ngalkuliyangku wantima, mai kuka wali-wali ngalkuwiyangku.'	'No, don't eat that, leave it alone. You don't eat plant food and meat together.'
'Nyaakun kukangka tjungura ngalkuni?'	'Why ever would you mix it with meat?'
Alatjingara ngalkupai, paura.	That's how it's eaten, after baking it.

wakati **inland pigweed**

maṟu-maṟu ***Portulaca oleracea***

A food source: the uprooted plants with mature, seed-laden capsules are allowed to dry out in a container to release the numerous, poppy-like seeds. These are then winnowed and yandied to remove the small capsule caps ***ipi*** (also the word for breasts). The cleaned seed is ground to a paste ***latja*** and small cakes ***nyuma***, ***wanytji*** are baked in hot sand with ashes and coals. The whole plant may also be baked and the root eaten.

Succulent, prostrate plant with reddish or brownish stems to 40 cm long, and a distinct taproot. Leaves fleshy to 2.5 cm long; flowers yellow, without stalks; fruit a capsule to 4 mm long, with a pointed cap that falls off to reveal numerous, minute, black seeds.

Occurs sporadically on sandplains, after rains.

Murika

Mai palunya yalkalpai, wirangka.	You thresh this plant in a ***wira*** dish.
Yalkaṟa(3) kaṉilpai. Kaṉiṟa waṉipai, ipi tjuṯa waṉipai.	Thresh it for a while and yandy it. After yandying it, you discard the 'breasts'.
Ipi mai panya paluṟu nyinanytja, ipi waṉinyi.	The breasts on the seeds. You discard the breasts.

*yalkani: threshing **wakati** (inland pigweed) in a **wira** dish*

Waṉira, kiliṉaṟa manta wiya. Pairpungkula kaṉini. Palunya, mai palunya rungkalpai, rungkaṟa...

Get rid of those and clean the seed of dirt. You yandy it by bouncing it up and down. Then you grind it up.

Mai kalkaṉi, mai palunya rungkaṟa ngalkuṉi.

First you separate out the seed, the food. Then you grind and eat it.

Pairpungkula kaṉiṟa, munu kaṉiṟa rungkaṟa ngalkuṉi.

You yandy it by bouncing it up and down; and after grinding it, you eat it.

Palunya kaṉiṟa rungkaṟa 'Wanyuṉa nyangatja uḻupunganyi.'

As you're yandying and grinding, you might be thinking, 'I'll just make this into flour.'

Palunya uḻupungkula wirangka pulapananyi.

You make it into flour, filling up a **wira** dish.

Munu paluṟu wanytjitjunanyi, tampa puṟiny panya.

And then you make a seed cake, like a damper.

Tampa pauṟa, nyinara(2) kaṯantaṟa wiṟuṟa kutu ngalkuṉi.

After baking the damper, you wait a while, then break it into pieces and eat it, really relishing it.

Mai wiṟu. Wanka titutjara nyinapai — kuṉtjul wiya, kilina...

It's fine food. We were always healthy — without colds, and clean.

Ungka ngalkupaitu. Nganana tjitjingku ungka warpulpai, munu paura ngalkupai... Katantarala paulpai.

We eat the root too. As children, we'd pull up the root, and bake and eat it. We'd break it off and bake it.

Nganana mai munuru ngalkula(2) pulkaringu.

We grew up eating many and varied plant foods.

***pairpunganyi**: yandying with a bouncing motion*

Vines, fungi and mistletoes

ipi-ipi — **caustic vine**

Sarcostemma viminale var. *australe*

A medicinal ***ngangkaṟi*** vine used in giving smoke treatment (fumigation) ***puyutjunanyi***: it is put on a fire to produce smoke which a sick person lies in for relief. Also used to make white decorative spots on skin.

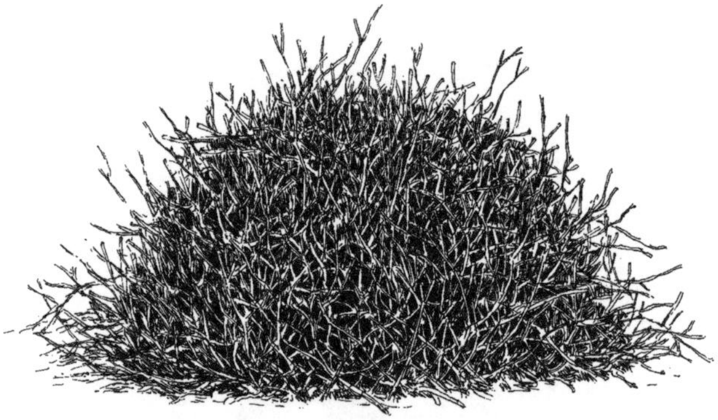

A sprawling, leafless creeper twining over rocks and plants. Stems grey-green, cylindrical, jointed and fleshy, exuding a milky sap when broken; flowers small, stalked, arising from joints; pods to 11 cm long, splitting longitudinally to reveal small seeds with a tuft of silky hairs.

Mainly found in stony soils on rocky hills.

Pompey Everard

Nyanga paluṟu atuṟa tjunkunytja wiya.	This plant isn't chopped up and put on the body.
Waṟu kutjaṟa unngu puḻka kutu kutjaṟa ka puyu puḻka pakalpai.	Instead you light a fire, and burn a whole lot of it beneath the fire, so thick smoke comes up.
Ka palula ngaripai, pika puḻkatjara.	And the sick person lies in it.

Ngarira(6) kampa kutjuparira ngarima piyuku.

After lying there for a long time, he'll roll over and lie on the other side.

Munu piyuku watungarama, watungarala(3) tjularingkula(2) itjan paluru tjulani kutu.

Then he'll lie face down. After lying on his front for a while, the stiffness goes — it really softens up stiffness.

Alatjingarampa pakaralta nyinama, munu kulinilta: 'Au, ngayulunatju kutu palyaringu!'

After this, he sits up and thinks: 'Hey, I'm really better!'

Puyu pulkangka watungarala(2) munu kankara alatji ngarira(3) kulini:

You lie in the thick smoke — face down, and on your back, and after a while think:

'Nyanga kutu pika palya panya. Wiyaringu kutu. Manararingu.'

'The pain around here has come good. It's all gone. It's gone numb.'

Ma-pulkaringu. Munu paluru watjani, 'Uwa, kanatju kutu nyaariwa, ipilyari.'

The sick person is content again. He'll say, 'Yes indeed, I'm really back in good shape.'

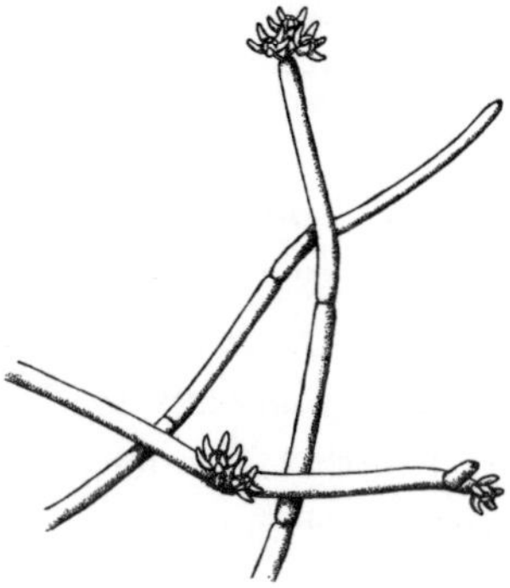

parka-parka

mistletoes

nga<u>n</u>tja

Lysiana murrayi
Lysiana exocarpi

Mistletoe fruits of both species are eaten raw. They are sweet and fleshy, with a sticky seed that is discarded. Often referred to as children's food ***tjitjiku mai***.

Mistletoes are stem parasites that attach to their hosts (often *Acacia* and *Eucalyptus* species) by a single swollen base. They differ as follows:

Parka-parka *Lysiana murrayi:* Leaves almost cylindrical, to 6 cm long, slightly pointed at the tip; flowers pink to red, greenish-white towards the base, borne on stalks to 2 cm long, singly or in pairs; fruit globular green at first, ripening to yellow, then to red or crimson.

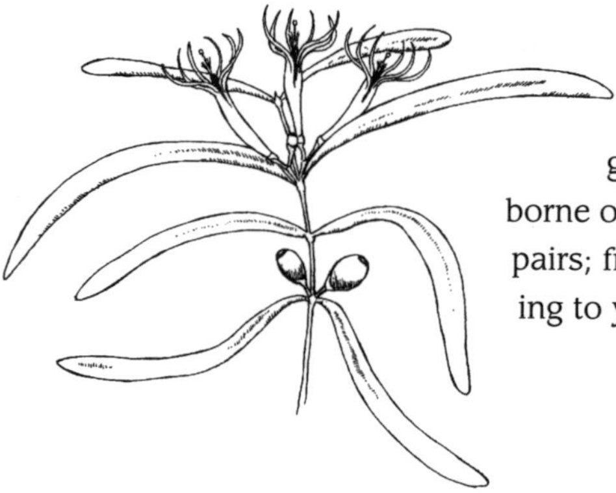

Nga<u>n</u>tja *Lysiana exocarpi:* Leaves opposite or sometimes clustered, smooth, flat, to 15 cm long and 1 cm wide, not thick, often curved; flowers in pairs or threes on a common stalk to 1.5 cm long, each flower on a further slender stalk to 0.5 cm long, yellow or red at the base, green at the tip; fruit ovoid, blunt-ended, green at first, ripening to red or black.

Ka ngalkupai tjitjingku, wiṟuṟa ngalkupai.	Children eat these. They really like them.
'Ngunytju! Ngayuku parka-parka kaṯantaṟa!'	'Mummy! Break off some ***parka-parka*** mistletoe for me!'
Ka parka-parka kaṯantaṟa(2) yungkupai.	So she breaks some off and gives it to him.
'Ka kutjupa nyangatja ngaṉtja!'	'And also some of that ***ngantja*** mistletoe here!'
Ka ngaṉtja kutjupa kaṯantaṟa tjunkupai. Palunya pulanya tjunkupai.	So she breaks off some of the other mistletoe and puts it down, puts the two of them down.
Ka ngura ngalkula(4). Ka piyuku puṉu kutjupa katima, kaṯantaṟa(2). Tjunkula kuka tjawanma.	And the child just eats and eats it. And she breaks off some more still and brings it. Puts it in front of the child and goes off to dig out some small game.
'Nyinama puṯa ngalkula, nyangangka, wiltjangka nyanga.'	'What about you just sit here eating, in the shade?'
Ka ngura ngalkula(3) panya palunya, tjitji paluṟu kunkunari.	And the child just eats and eats that stuff, and falls asleep.
Ka palumpa ngunytjungku tjawaṟa(3) kuka ma-witila!	And his mother can be off digging out some game!

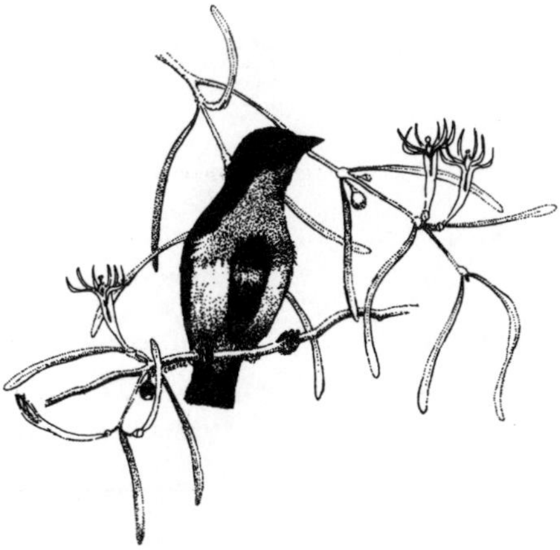

urtjan(pa)

spear bush

Pandorea doratoxylon

The stems are used as hunting spear shafts ***katji muṟu***. They are cut from the bush, stripped of bark, warmed over a fire and then straightened by bending. A blade ***waṯa*** and a barb ***mukul(pa)*** of mulga are affixed with kangaroo sinew ***pulyku***, ***marpany(pa)*** and resin ***kiṯi***.

A distinctive shrub to 4 m tall, with many arching and intertwining vine-like stems. Narrow leaflets taper to a point and are arranged in opposite pairs; scented flowers are cream coloured, tubular with brown–purple markings in the throat; capsules ***maṟa*** are elliptical, beaked and split open lengthwise to release many papery, winged seeds.

An occasional species of rocky gorges, gullies, and sheltered hillsides among rocks.

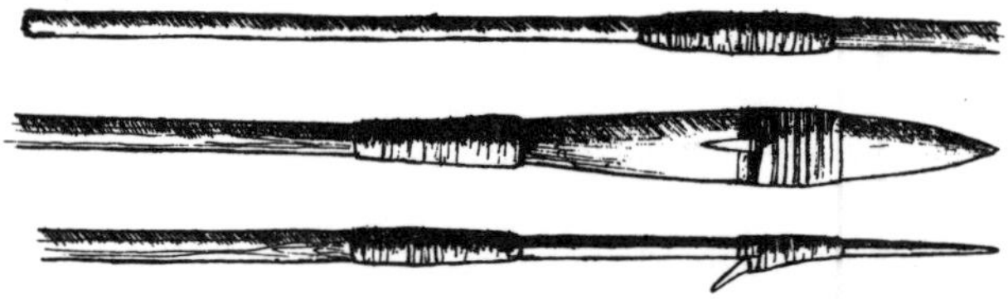

Kanytji

Katji nyangatja waṯuṟa tjukaṟuruṟa.	You straighten the spear-vine by bending it.
Kulpara, waṯuṟa tjukaṟuruṟa, winarpungkula(2), tjunkula waṉikatinyi.	After going back to camp, you bend it straight, notch the ends to joint it, and then put it aside.
Munu yankula katji waṯa atuṉi.	Then you go and chop out the spear blade.

Atu̲ra katira, palu̲ru nyangatja iri̲ra̲, winarpunganyi.	After bringing it back, you sharpen it, and notch it to joint it.
Munu mat̲i-mat̲ipungkulampa, pulykulta ngalkun̲i, nyulytjar̲a, yur̲ur̲ampa.	Then after scoring the joint, you chew some kangaroo sinew, to soften it.
Kit̲i, kutjar̲a(2), pungku-pungkula kit̲i kun̲tjir̲a, ngat̲i nyanga, munu wantinyi.	Then heat some resin and sort of tap it on, smear the resin on the notch and leave it for a while.
Munu palunya ngapartji, wat̲a ngapartji kit̲ingka kun̲tji-kun̲tjir̲a, wantinyi.	Then you lightly smear the blade with resin too, and leave it for a while.
Munu wat̲a, katji panya witini munu war̲ungka ngilytjitjur̲a, yur̲uringkunytjalampa, ngampaltjunanyilta.	Then you take hold of the blade, and the spear, and after holding them in the warmth of the fire — once the resin goes soft, you stick them together.
Ma-tjungur̲ampa witangkalta kun̲tji-kun̲tjini.	As you join them together you lightly smear it with spittle.
Witangka kun̲tji-kun̲tjir̲ampa, aa pil̲t̲iringkunytjitjangkampa, kantiringkunytjitjangkampa,	Once it's smeared with spittle, and after it's gone dry, after it's gone firm,
katji nyanga an̲angu kutulta war̲ungkalta tjuna-tjunanyi.	you quickly pass the body of the spear through the fire a few times.
Ka yur̲uringkula katu nyanga ka lipulananyilta.	So the resin at the top gets soft and spreads out.
Unytjungku lipular̲ampa, mantangka kun̲tji-kun̲tjira, kit̲i pil̲t̲iringkunytjitjampa, pulykungkampa panya karpir̲a wantinyi.	Once you've made it spread out just a bit, after smearing it lightly with earth, once the resin has gone dry, you bind it again with kangaroo sinew, and put it aside.
Munu ngarira tjin̲turingkulampa mungawinki piyuku kutja-kutjan̲i.	After sleeping the night, you lightly heat it again the next morning.
Kutjar̲a, yur̲ar̲a ka kit̲i nyanga paka-pakalaran̲i, marpanytja ngur̲ur kutjarawanu.	Heat it and soften it, and the resin comes up here, through the sinews.

Ka witira, marungka witira, witangka kuntji-kuntjirampa, piltiringanyilta.

And after wetting it with your hand, to smear it all over with spittle, it goes dry.

Ka warungka watura, mira-mirara tjukarurura mulatu, nyakula.

Then you flex it over the fire, watching it closely as you straighten it perfectly, watching it.

Tjinguru tjukaruru katjingka lipula ngarama wata nyara ka wantinyilta munu karpinilta.

Maybe the blade'll be straight on the spear. If so, you leave it for a while, and then bind it up more.

Karpirampa piyuku tjunkula wantinyi.

After binding it, you put it aside again for a while.

Palu mukuringkulampa, mukul palyani.

Of course if you want, you can make a barb.

Mukul irirampa, wata karpira wantinyi, munu mukultjunanyi, palulakalta. Mukultjura, nanngu panyalta tjunkula wantinyi.

After sharpening it, you bind it onto the blade and leave it. The barb goes on. Then after fitting the barb, you put the spear aside for a while.

Munu tjunpangkalta warulyarulyinanyi, tjunkula wantinyi.

And then afterwards make it nice and warm over the hot ashes, and put it aside again.

Munu pal̲ka ngapartji kutja-kutjar̲a karpini, munu karpir̲ampa, panya witu ngar̲anytjaku, kanti-kanti panya ngar̲anytjaku —

After that you turn to the end section, heating it up, and binding it on, and after binding it on — you know, so it'll be secure, so it'll be really firm —

alatjingar̲ampa nyakula tjukar̲urur̲a — kal̲i-kal̲i nyanga pur̲iny nyanga ipangkalta wat̲un̲i.

you straighten it out like this, watching it — a crooked bit like this you bend out over the ashes.

Ipangka wat̲ur̲a wan̲ar̲a, tjukar̲urur̲a, munu tjukar̲urur̲ampa, tjunkula wantinyi.

You flex it all along, straightening it, and once it's straight, you put it aside.

Munu pal̲ka nyanga tjukar̲urun̲it̲u. Pal̲ka wat̲ur̲a tjukar̲urur̲a wantinyi.

And you get the end section here straight just the same. Straighten it out by bending it, and put it aside.

Munu kutjupa ngapartji palur̲u pur̲iny. Nyangatja winarpunganyi, ala nyangatja, panya alatji,

And in other cases, just the same, you do this. You'd have to notch this one just here, for instance,

nyanga kal̲i-kal̲i nyangatja tukul ngar̲anyi, ka ala nyangakutu, ala nyangakutu winarpunganyi, tjan̲angka nyanga. Alatji kutu ma-ir̲ir̲a.

because there's a noticeable crooked bit. So you sort of notch it here — on the back of where the bump is. You shave it back around here.

Wai ir̲ini pul̲kar̲a. Palu tjukutjuku ngat̲ir̲a wantinyi.

Of course, you don't shave it much. Just make a small hollow.

Munu wat̲ur̲a tjukar̲urur̲ampa pulyku nyanga alatjingar̲a karpir̲ampa, nyangatjalta wat̲un̲i munu tjukar̲urun̲ilta.

And after bending the crooked bit out, you bind the spear around here with sinew, bend it and straighten it.

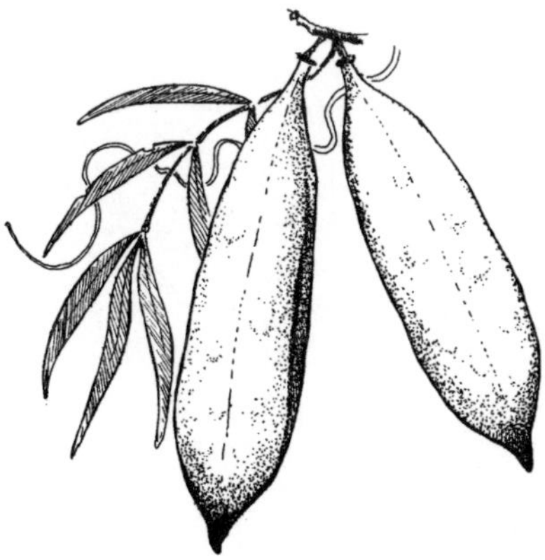

Munu alatji tjunkulampa nyakula mulampa, alatjingar̠a. Kampa kutjupananyi.

Then you put it out like this, and have a really close look at it. You turn it around and around.

Munu alatjir̠ampa nyangatjalta ngapartji ngat̠ini.

Because you have to shave it back further down on the other side, to balance it.

Ngat̠ir̠ampa, katji pal̠ka ngapartji urtjan. Urtjanpat̠u ngat̠ini, munu tjunanyi.

Then there's the end section, made of spear-vine. You make a depression in the end, and fit it on just the same way.

Munu ngura kit̠ingka kutja-kutjar̠ampa, yur̠u-yur̠ur̠ampa kutjaraka tjunanyi.

Just heat up some resin, to soften it, and put it on the two pieces.

Munu war̠ungka ngilytjitjur̠ampa panya palur̠u pur̠inykalta ngampaltjunanyi.

Put them in the warmth of the fire, and stick them together just the same way.

Munu ngura mar̠angku witir̠a tjukar̠urar̠a mula, lipula mula nyakula...karpiniltar!

And after gripping it with your hand so it's on really straight — once you see it's on straight — you bind it on!

Grasses

kaḻṯu-kaḻṯu — native millet

kutja — *Panicum decompositum*

A food source: the seeds are gathered in a ***mimpu*** bowl by rubbing heads off the stalks, then winnowed, yandied, moistened, ground to a paste, and baked as damper ***nyuma***, ***wanytji*** in hot sand and ashes.

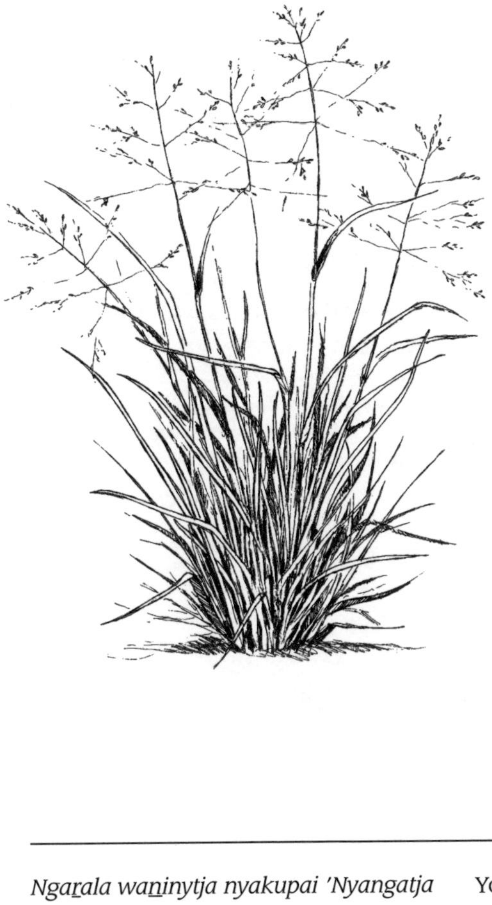

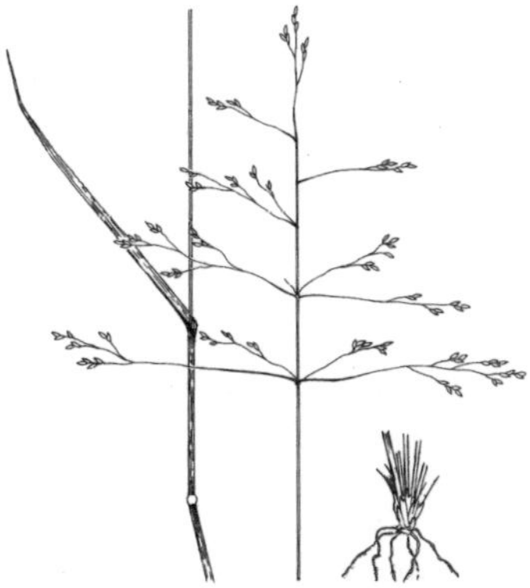

Stout, tussock-forming perennial to 1 m high. Stems hollow, erect; leaf blades to 1–2 cm wide, flat; seedhead to 40 cm long, and as wide, with relatively few spikelets.

Most common on the banks and floodouts of watercourses. Also in grasslands on sandplains. Seed ripening after summer rains.

Mollie Everard

Ngaṟala waṉinytja nyakupai 'Nyangatja puḻkanya ngaṟanyi, kaḻṯu-kaḻṯumpa!'	You'd see a lot scattered around. 'There's plenty here, native millet!'
Ka nyulkuṟa, warpuṟa tjunkula(5) mimpungka.	We'd rub the seeds off, after pulling the plants out and putting them in ***mimpu*** bowls.

Warpura tjunkula(8) tjunkulampa nyulkura tjunama.	We'd pull up lots of plants and put them in the ***mimpu*** bowls.
Nyulkura wanira punu panya, nyulkura wanima(2).	After rubbing the seeds off each plant, we'd throw it away, one after another.
Munu pulkara katima(4), nyulkura katima.	We'd build up a big lot of seed, taking it along with us, as we went.
Munu kalkara mulatu katipai, kankukutulta.	And after really getting out a lot of seed, we'd take it to the shade.
Marangkula nyulkulpai. Nyulkura(2) ararangkula(4) tjunama.	We used to rub the seeds off by hand. Rub, and then winnow it for a while before putting it aside.
Kapingka kulya-kulyara, mara kulya-kulyara kapingka, nyulkulpai,	We'd sprinkle our hands with water, to rub it out.
Nyulkura(2) ararangkula(2) kalkara mulalta.	After rubbing it out and winnowing it to extract the seed alone, we'd...
Ngura, ka kalka mulalta tjunkupai, katipai ngurakutu.	Well, we'd just have the clean seed left, and take it to camp.
Kapingkalta kulya-kulyara, wiita tjunkupai, mai paluru tjularingkunytjaku.	There we'd moisten it with water, so it'd go soft.
Munu tjulalta rungkara(2) ngurangka ngalkuni. Palya.	Then we'd grind up the soft seed, and eat it. All right.

ararangany*i: winnowing seed from chaff*

kutanu **greybeard grass**

Amphipogon caricinus

This grass can be used as a hat (sometimes moistened first), or be made into a circular head pad ***manguri*** for carrying things. Also used as a strainer when drinking water from waterholes, and honey drinks from vessels; as a sponge to take up water from crevices in rock; and in making poultices with medicinal herbs, such as ***kalpipila***.

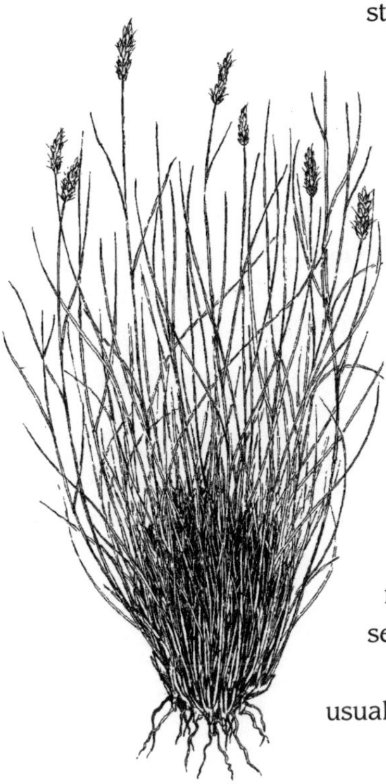

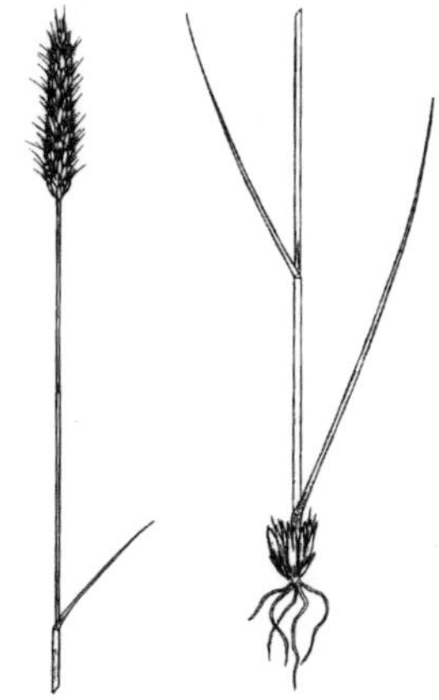

An erect, perennial, tufted grass to about 60 cm high, with numerous wiry stems. Leaves narrow with inrolled edges, stiffly pointed; seedhead cylindrical up to 6.5 cm long. Occurs sporadically in a variety of habitats, usually on sandy soils in association with spinifex.

Murika

Ngayulu kutanu wangkanyi.	I'll talk about greybeard grass.
Nyanga palunya yannga kulpara(2), kutanu warpulpai.	You pull up some of this grass while travelling, on a trip.
Munu paluru kapiku kulpayinkupai munu kulpara nyakunyakukatipai.	And bring it back for water. You keep an eye out on the way back.
Tjarpara, paluru kutanu palunya tjunkupai.	After arriving you put the greybeard grass down.

Munu kapilta tjikira kutanu palunya katangka tjukula. Kulya-kulyara, kata karpira kulpapai.

And having drunk some water, you put the greybeard grass on your head. After sprinkling yourself with water, you'd tie the grass on the head and go back.

Munu kulpara, kankungka ma-tjunkula, kuka paura ngalkupai.

And after returning, you set up in the shade, then roast and eat some meat.

Nyanga palunya kunyu kuka kalaya wanantjikitjangku kutanu nyanga palunya tjunkula, wanalpai. Iriti.

They say that if you wanted to chase an emu, you would put this grass on and go after it. In the old days.

Munu wanara ma-pungkupai, kalaya. Munu kulpara watjara iyalpai.

They would chase and kill an emu. Then after returning to camp, send someone off.

Ka paluru yankula kapitjarangku uralpai. Kuka palunya paura(2) mungartji katipai.

And he would go, with water, and get it. Roast up the meat and carry it back in the late afternoon.

Nyanga paluru miri tjutaku kutanu. Anangu winkiku. Tjikilpai, kapingka tjunkula.

This grass is from the people who have passed away. It's for all Aboriginal people. You can use it when you're drinking, by putting it in the water, as a strainer.

Mukata wiyangka, nyanga palula kutanu tjunkula katipai.

Before there were hats they would carry things wearing this grass.

*A **mimpu** bowl being carried with the aid of a headpad **manguri***

wanguṉu

naked woollybutt

Eragrostis eriopoda

A food source: the seedheads are rubbed off, and the chaff is separated from the seeds by pounding, singeing, winnowing and yandying. The seeds are then ground to a flour, and water is added to produce a runny mix, which is cooked in hot sand, ashes and coals to produce a traditional bread or damper ***nyuma***, ***wanytji***.

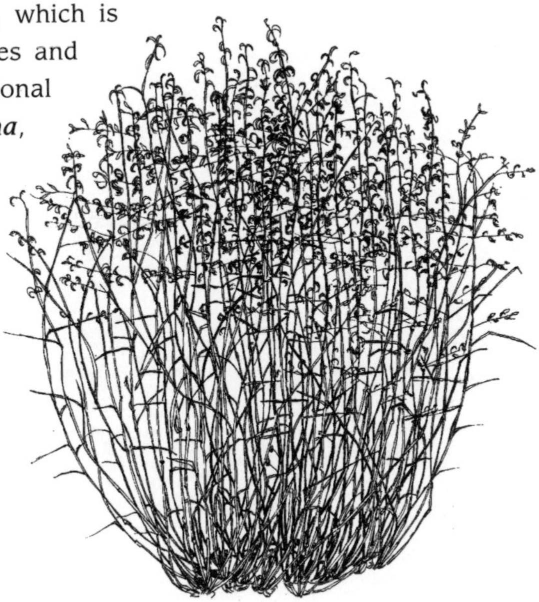

A perennial, tussock-forming grass to 60 cm high with coarse, hairy roots, and a dense, woolly butt. Stems erect and wiry; leaves narrow, with in-rolled edges and rigid points; seed-head to 20 x 7 cm, tapering to a point, with numerous spikelets.

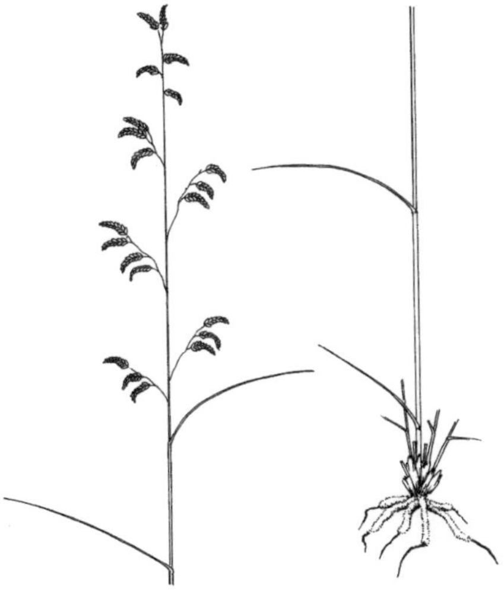

Widespread, mainly on spinifex sand-plains and sandhills, and in mulga woodland. Also on river floodplains and levees, and on limestone rises. Seed ripening in autumn or winter after summer rains.

Mai panya wangun̲u nyut̲un̲i. Nyut̲ur̲a(2), kan̲ini.

You rub out the woollybutt seed. Separate the seed by rubbing, and then yandy it.

Kan̲ir̲a tir̲alpa maunta tjunanyi. Mai maunta tjunanyi, initjara tir̲alpa.

After yandying it, you put a separate portion aside, a reserve portion. It has the name ***tiralpa***.

Munu palulangur̲u rungkar̲a mai palunya ul̲upunganyi.

After that you grind it up into flour.

Ul̲upungkula wirangka tapaltjur̲a, ul̲upungkula pul̲kar̲a, mai palunya,

You grind it, after putting a ***wira*** dish at the back of the grindstone, to collect the flour. Make a whole lot of flour,

kapingka kulya-kulyan̲i, kulya-kulyar̲a nyumatjur̲a, tampa pul̲ka nyumatjunanyi.

and then sprinkle it with water. Moisten it and make a seedcake, a big damper.

Munu palulangur̲u kat̲ar̲a — tjitjingku, watingku, minymangku tjitjingku ngalkun̲i.

After that you cut it up and the children, men and women all eat.

Ngalkula(2) ngarinyi, ngarira tjin̲t̲uringanyi.

After a good feed, they sleep the night.

***Witan̲i**: singeing the seed heads with burning twigs*

Munu ngarira tjinturingkula pakara nyutuni, nyutura(3) kantura(6).

Having slept the night, they get up and start rubbing seed out again — rubbing and stamping it out.

Mai palunya kanini, tiralpa tjunanyi mauntalta, tjintungka rungkara ngalkuntjikitjangku, tjintukutu.

You see, you yandy it and put a reserve portion aside, to grind and eat the next day — for the next day.

Munu mai palunya mungartjitja rungkara ngalkuni. Uwankara rungkani.

So then you grind up and eat the stuff from yesterday. Grind up the lot.

Tjitji, paluru tjitji pulkani, nyiinkananyi, pulkani tjitji.

Thus a mother 'grows up' a child, makes him a big boy (a ***nyiinka***).

witita — native truffle

Choiromyces aboriginus

This much-relished fungus is dug up from just beneath the ground, where the mature fruiting bodies cause cracks. They are cooked in hot sand, ashes and coals and eaten.

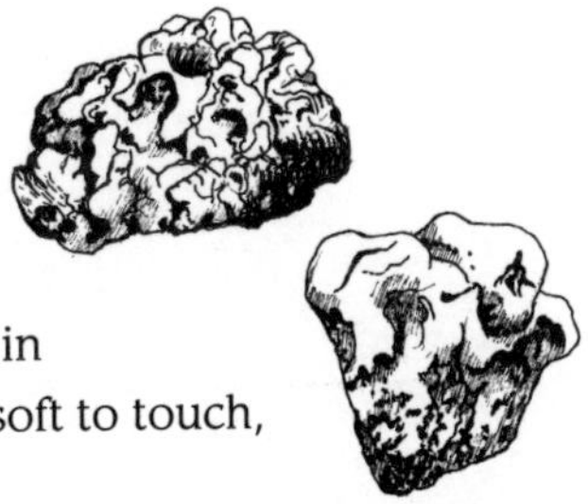

An amorphous fungus which varies in size from just a few centimetres to 10 centimetres in length (from the tapered base to the top). It is soft to touch, and will break up easily into pieces.

Found mainly in woodland/shrubland, on sandplains and sandhill country, and at the base of rocky hills, after adequate rains.

Ka puṯingka tjarpama. Nyaa ngaṟapai, mai kutjupa?	Say you go into woodlands. What grows there?
Mai paluṟu kutjupa itjaṉu puḻkangka ngaṟapai.	A different type of food plant that grows in lush green country.
Witiṯa, mai witiṯa. Paluṟu tulyku puḻkangka ngaṟapai, witiṯa paṉya. Apu panta nyangangka. Ngura tjuṯangka.	The native truffle. It grows in muddy areas. Near the sides of hills. In lots of places.
Ngangangka ngaṟanytja wiya. Nyara apuwanu, nyarawanu kutju ngarapai. Apu nyarangka nyara.	It doesn't grow here. But over there in the hills. It only grows around there, in those hills there.
Witiṯa palunya paulpai, kapingka yuṟuṟa.	They used to bake the truffle, after moistening it in water.
Uu wantipai, nyara palunya. Wari puḻka ngalkupai, ngalkula nyaaringkuku? Nyurkaringkupaingka.	Umm, and leave it, to eat it really cold — in case you get skinny.
Tjinguṟu nyurkaringkuku witiṯa tjuṯa kutju ngalkula. Waṟu ngalkula, paluṟu nyaa? Nyurkaringanyilta.	You can get skinny from eating just truffles. If you eat it too hot, you get skinny.

GLOSSARIES

Please bear in mind that none of these lists is complete, and that there are certain to be some inaccuracies. It should also be noted that these words are specifically Yankunytjatjara usage, as at Mimili. Pitjantjatjara usage may differ in some cases. Finally, some of the words listed have more general meanings than those indicated here, which are restricted to the context of plants, and plant use. Further information may be found in IAD Press's *Pitjantjatjara/Yankunytjatjara to English Dictionary*.

Plant parts and related terms

anangu	(lit. 'body') trunk and stems of a plant, excluding leaves
aparuma	edible sweet crusty scale on the leaves of the river red gum ***apara*** and other eucalypts
atjirikitikiti	type of leaf gall on mulga ***kurku***
ipi	(lit. 'breast') small cap or projection on fruits or seeds, e.g. inland pigweed ***wakati***
itirki	type of inedible insect gall on stem
iwiri	lateral, branching root
kaliny-kaliny(pa)	spiked nectar-bearing flower of corkwood ***witjinti*** and some grevilleas, e.g. ***ultukun(pa)***
kalka	(lit. 'hard, roundish thing') seed, bulb, fruit
kalpi	broad leaf (also feather)
kalpi-kalpi	pod of *Acacia* and *Senna* (*Cassia*) species
kanpa	poisonous, inedible
kiti	resin, adhesive gum
kultu	(lit. 'upper part of body') trunk
kuru	(lit. 'eye') ripe
kumpu	(lit. 'piss') yellowish acrid, bitter juice in some bush tomatoes, e.g. ***ituny(pa)***

kurku	honeydew, 'mulga honey' — sweet liquid sometimes found on branches and stems of mulga, produced by sap-sucking insect parasites, known as scales; also a general term for mulga trees
kurkunytjungu	honeydew, 'mulga honey' (as above)
likara	bark
mai	vegetable food; food plant
makaly(pa)	hollow branch
mangir(pa)	dried out reddish lumps found on the branches of mulga ***kurku***, made by a scale insect (see ***kurku***)
mara	(lit. 'hand') finger-like capsules, e.g. of corkwoods ***witjinti***, grevilleas and spear bush ***urtjan(pa)***
mina	(lit. 'arm') branching stem, branch
mirpur(pa)	bole, a hemispherical swelling on the trunk of river red gum ***apara***
mulya	(lit. 'nose') dried but reconstitutable fruit, e.g. wild fig ***ili***, quandong ***mangata***
muntu	green, unripe
murtjul(pa)	(lit. 'knot', 'tangle') leaf gall, e.g. as on colony wattle ***tjuntala***; a knot in timber
ngampul -ngampul(pa)	small, inedible fruit or berry; hard bud
ngapari	edible sweet crusty scale on the leaves of river red gum ***apara*** and other eucalypts
parka	narrow leaf; broken-off branches with narrow leaves, used in butchering game, for example, to keep meat clean
punu	plant, especially trees and shrubs; stick or piece of wood; thing made of wood, e.g. wooden tool
pura	unripe, green
putja	grass
talytja	small offshoot or fork
tarka	(lit. 'bone') woody stem, e.g. of native tobacco ***pulyantu***, and bush banana fruit ***utiralya***
tatu	inedible fruit of eucalypts, native willow ***kumpaly(pa)***; stone of quandong ***mangata***

tjarapakutjara	fork in tree
tjau	edible gum
tjiily(pa)	taproot, e.g. of mulga ***kurku***
tjilka	prickle, thorn; prickly, thorny
tjinytjulu	small gumnut, specifically that used in hair decoration
tjuratja	nectar, honeydew and other sweet substances, such as scales
ukiri	green plants, green grass, herbs; specifically used as a term for native tobacco ***pulyantu***, ***mingkul(pa)***
ularama	flower of corkwoods ***witjinti***
ungka	taproot of herb or succulent
uniny(pa)	(lit. 'hard roundish thing') seed, bulb, fruit
unturuntu	flower
unytjunytju	fallen leaves, leaf litter; husks and chaff from winnowing etc.
wama	nectar, honeydew, and other sweet substances, such as scales
wata	base of tree trunk, upper part of root
yultu	hollow (e.g. trunk)
yuti	aril, i.e. seed appendage

Habitats and Vegetation

apu	(lit. 'rock', 'stone') rocky outcrops, hills, ranges
arpata	limestone, limestone rises
itjanu	lush country
karu	watercourse: gully, creek, riverbed
manta	ground, earth
pila	spinifex grassland, plain
puti	woodland, shrubland, scrub, bush

putu	old flattened termite nest, making patch of very hard ground that is often used in threshing and cleaning seed
tali	sandhill, sand-dune, sandhill country
tjanpi	spinifex grass, grass plain
tjata	dense growth, patch, thicket, stand, grove
tjintjira	claypan, swamp
tjulpir(pa)	mud
uril(pa)	open country, plain, clearing

Processing plants
(stages and products, processes and implements)

Note that wood-working, spear-making terms and so on are not included. Verbs are given in their present tense form.

araranganyi	winnow
atuni	pound with stone, chop, mash
kalkani	extract the grain or fruit
kampanyi	burn, cook
kanil(pa)	shallow wooden dish, used as a yandying vessel
kanini	yandy, i.e. shake back and forth in a shallow dish, e.g. to separate seed from chaff
kantuni	stamp (also kick, dance with stamping step)
kaputu	lump, wad, poultice, quid, 'ball'
kaputunanyi	make into a lump (poultice, quid etc.)
karpini	bind on, tie up
katani	cut
katantananyi	break
katawakani	cut by piercing
kulya-kulyani	sprinkle, put water on
kuntjini	rub on/in, e.g. rub grease, paste into skin

kurapunganyi	rub desert raisin fruit on ground or between the palms, to bruise it
kutjani	heat, boil; make fire
lampini	peel, prise off, lever
latja	a paste, mash
latjapunganyi	make into a paste, e.g. by grinding with water
lungkun(pa)	viscous yellowish paste, e.g. seed paste, egg yolk, inside of witchetty grub
mangkalpai	apply to the hair ***mangka***
milyintji	husks, removed by winnowing
mirka	food stored for later
mimpu	hemispherical wooden bowl, used to carry water
mimpungkananyi	put into a ***mimpu***
minani	suck liquid off something
ngangkarinanyi	heal, act as healing agent
nganytjawali	soft e.g. steamed greens, such as ***unmuta***
nyitini	rub grease, fat or ointment into something
nyulkuni	rub (grass, pods) between the palms to free seed
nyuma	traditional bread, damper or cake of ground seed and water, for baking
nyumatjunanyi	make into a damper
nyutuni	rub (grass, pods) between palms to free seed
patapatani	shake (so that, for example, dust or fruits fall off)
pairpunganyi	yandy with a bouncing action
pauni	bake, roast, parch, scorch; burn design onto wood
piti	large, hemispherical wooden bowl, used to carry water
pulparu	sloppy paste of ground grass seed, e.g. of woollybutt ***wangunu***
riwipunganyi	swing back and forth, e.g. to wave burning leaves, bark, fire-stick so it burns better

ruku	leftover food, on grindstones or dishes. Generally, a visible reminder of anything
rungkani	hit with hurled missile; grind, file; knead
runyu	mash, pulp
runyuni	mash, pulp up
tangka	firm, set, congealed
tapal(pa)	covering; a vessel such as a ***wira***, used to collect flour from a grindstone
tapaltjunanyi	put ***wira*** as collecting dish while grinding seed
tiral(pa)	portion of grain put aside to be ground later
tjiwa	flat stone; lower stone of two grindstones
tjulani	soften
tjulkuni	squeeze juice ***tjulku*** out of something
tjungari	round, upper hand-held grinding stone, millstone
tjunyini	knead in water, squish around in water (to dissolve)
tjutini	get or pour water
tjuultjunanyi	make a pile of something; heap something up into a pile
tjuuntjunanyi	pile dirt on something
ulu	powder, flour
ulupunganyi	grind or crush to flour, powder
urkuni	pull out, uproot
utulunanyi	gather together, form into a group ***utulu***
wakani	pierce, spear, sew
waluni	bind or fit on tightly ***walu***
waltuni	cover
wana	digging stick
waninyi	discard, throw down
wanytji	traditional bread, damper or cake of ground seed and water, for baking
wanytjitjunanyi	make into a damper

warnyun̲i	rip off, e.g. stripping leaves or fruit from stem
warpun̲i	pull up, rip out
wira	small cup-like wooden dish used for digging, drinking and collecting berries
witan̲i	singe, burn in flames, fry
yalkan̲i	thresh, pound with stick
yur̲un̲i	make into liquid ***yur̲u***
yuun̲tjunanyi	bundle up, make into a bundle or tangle ***yuun̲(pa)***
yuwun̲i	rock back and forth, e.g. a dried-out tree, to loosen

Annotated list of additional plants

in alphabetical order of Yankunytjatjara name

Yankunytjatjara name	Common English name • Usage/Notes	Botanical name	Descriptive term
altar(pa)	gum-barked coolibah	*Eucalyptus intertexta* *Eucalyptus sparsa*	mallee
anultja	bindweed • used to bundle up edible greens, e.g. ***unmuṯa*** native cress, for cooking in warm earth and ashes	*Convolvulus erubescens*	morning glory creeper
apita	garland lily	*Calostemma purpureum*	lily
aratja	native fuchsia	*Eremophila rotundifolia*	shrub
aripita	cassia (senna)	*Senna artemisioides* ssp. *helmsii*	shrub
awalyuru pakaḻi-pakaḻi waṉakaṯakaṯa	bush currant • black, sweet currant-like fruits eaten raw, or pulped with water	*Canthium lineare*	tall, slender shrub
ikaṯuka	—	*Acacia calcicola*	tree
iḻi	wild fig • an important food **mai puḻka**: the red, ripe fruits are eaten raw; the fallen dry fruit may be ground with water to make an edible paste, or made into balls for later use	*Ficus brachypoda*	spreading tree
ilintji	scent grass umbrella grass silky brown top kangaroo grass	*Cymbopogon ambiguus* *Enteropogon acicularis* *Eulalia aurea* *Themeda australis*	tall grass
ilpaṯilpaṯa mamawaṟa tjilu	stalked puffball • black powder used by men and boys to darken facial and body hair	*Podaxis pistillaris*	fungus

ilpili	ti-tree, inland paperbark	*Melaleuca glomerata*	small tree
	• stems used for children's spears		
intiyanu	—	*Stemodia florulenta*	herb
	—	*Olearia stuartii*	subshrub
	apple bush	*Pterocaulon sphacelatum*	herb
	blue rod	*Stemodia viscosa*	herb
inturalkalpai wilypin-wilypin(pa)	—	*Spartothamnella teucriiflora*	leafless shrub
ipiri kara	curly wire grass	*Aristida contorta*	grass
	• made into a pouch to carry small fruits such as bush tomatoes		
iriya	old man saltbush	*Atriplex nummularia* var. *nummularia*	shrub
	bladder saltbush	*Atriplex vesicaria*	shrub
	silver bush	*Ptilotus obovatus* var. *obovatus*	subshrub
iriya tjilkala	rolypoly, buckbush	*Salsola tragus*	shrub
itara	bloodwood	*Corymbia opaca*	tree
	• the inedible fruits are called ***tatu*** • bloodwood apple ***angura*** is an edible insect gall; also called 'bush coconut', the woody gall is cracked open and the white flesh inside eaten		
iwatiwata malkakutjal(pa)	—	*Einadia nutans* ssp. *eremaea*	subshrub
iwatiwata malkakutjal(pa) mukul-mukul(pa) [***mukul(pa)*** 'hook', 'barb']	—	*Rhagodia eremaea*	spindly shrub
kaliwara	—	*Acacia olgana*	tall shrub
	• the seeds are gathered, winnowed, parched in hot ashes, yandied, ground with water to a paste and eaten • edible grubs ***maku*** are found in the roots		

kalpa<u>r</u>i	rats' tails	*Dysphania kalpari*	scented herb

- the ripe seed heads are gathered, singed, then winnowed, yandied, and ground into a meal (which can be mixed with honey from honeyants *tja<u>l</u>a*), baked and eaten
- used to make a medicinal wash for skin rashes

ka<u>l</u>pil-ka<u>l</u>pil(pa)	—	*Einadia nutans*	vine
		Glycine canescens	vine

ka<u>l</u>pil-ka<u>l</u>pil(pa) papawiti<u>l</u>(pa)	—	*Mukia maderaspatana*	creeper vine

- medicinal: mashed to make a poultice or rubbing ointment

ka<u>l</u>pilya kurku	mulga (long, narrow, flat leaf-form A)	*Acacia aneura*	tree

- wood used for making boomerangs **ka<u>l</u>i** spear throwers **miru** and other items requiring hardwood

ka<u>l</u>pir-ka<u>l</u>pir(pa)	Sturt's desert rose	*Gossypium sturtianum*	shrub

ka<u>l</u>u<u>t</u>i kantu<u>r</u>angu	desert poplar	*Codonocarpus cotinifolius*	small tree

- edible beetle larvae grubs, known as **maku kata<u>t</u>i**, are recovered from the lower trunk and roots by snapping the tree at the base

kampu<u>r</u>ara kampu<u>r</u>ar(pa)	desert raisin	*Solanum centrale*	subshrub

- fresh and dried fruits are eaten raw; dried fruits may be cooked, pounded, mixed with water, and the paste eaten or made into a ball for later use

kantu<u>r</u>angu	see **ka<u>l</u>u<u>t</u>i**
ka<u>r</u>a	see ***ipi<u>r</u>i***
karinga<u>n</u>a	see main section; see also ***intiyanu***; ***tjaamu<u>l</u>u<u>r</u>u***

karpil-karpil(pa)	cassia (senna)	*Cassia pleurocarpa*	low shrub
ku<u>l</u>i ku<u>l</u>ilypuru	native pine	*Callitris glaucophylla*	tree

kulypur(pa) tawal-tawal(pa)	wild gooseberry	*Solanum ellipticum*	subshrub
	• has edible fruits; Yankunytjatjara people distinguish between ***kulypur(pa)*** and ***tawal-tawal(pa)***, indicating that the former is found in sandhill country, and the latter in both sandy and rocky situations • ***tawal-tawal(pa)*** leaves may be used as a tobacco substitute when there is no ***ukiri*** wild tobacco around		
kumpaly(pa)	native willow	*Pittosporum angustifolium*	small tree
	• the fruits are considered poisonous		
kumpul(pa)	*see* ***ituny(pa)***, main section		
kurara	dead finish	*Acacia tetragonophylla*	shrub/small tree
	• seeds are gathered, winnowed, parched in hot sand and ashes, yandied, ground with water to a paste ***latja*** and eaten		
kurkara	desert oak	*Allocasuarina decaisneana*	tall tree
kurku	*see* ***wintalyka***, ***minyura***, main section; *see also* ***kalpilya***, ***puyukara***, ***tjamalya***		
kurumaru untalya	wheel fruit	*Gyrostemon ramulosus*	small tree
	• fruits called ***ngampul-ngampul(pa)***		
kutja	*see* ***kaltu-kaltu***		
liru-liru [***liru*** snake]	pussytails tall yellow top	*Ptilotus* species *Senecio magnificus*	subshrub subshrub
malkakutjal(pa)	*see* ***iwatiwata***		
mamawara	*see* ***ilpatilpata***		
mangka-mangka [***mangka*** hair]	matspurge, milk weed	*Euphorbia drummondii*	prostrate herb
	• food for rabbits and kangaroos, but poisonous to sheep • put on children's heads, to protect them from the sun		
mani-mani	a native fuchsia	*Eremophila paisleyi*	shrub

mintjingka ngarankura	a native fuchsia	*Eremophila latrobei*	shrub
	• nectar ***tjuratja***, ***wama*** is sucked from the individual flowers, once they have been plucked from the bush		
mintju	—	*Acacia melleodora*	tall shrub
mukul-mukul(pa)	*see* ***iwatiwata***		
munyun(pa)	red poverty bush	*Eremophila duttonii*	tall shrub
munyun(pa) watara	turpentine bush	*Eremophila sturtii*	tall shrub
	• used in the smoke treatment of backache, bad cough and general sickness		
muruntu-muruntu	—	*Dicrastylis costelloi*	woolly shrub
ngalta	desert kurrajong	*Brachychiton gregorii*	tree
ngapari tjintjulu	water mallee, red mallee	*Eucalyptus eucentrica*	mallee
ngarankura	*see* ***mintjingka***		
ngatun(pa)	prickly wattle, acacia bush	*Acacia victoriae* var. *arida*	tall shrub /small tree
	• ripe seeds are gathered, winnowed and yandied, parched in hot sand and ashes, cracked with a stone and ground with water to an edible paste • edible gum ***tjau***, a yellowish, crystalline, brittle type		
ngau-ngau	—	*Sida filiformis*	herb
nyinkini nyiruny(pa)	mistletoe	*Amyema maidenii*	pendulous parasite
	• the sticky edible fruit, concealed in a hairy grey covering, is considered a children's food		
nyiruny(pa)	*see* ***nyinkini***		
pakali-pakali	*see* ***awalyuru***		
pakuta palpa	horse mulga	*Acacia ramulosa*	tree
palpa	*see* ***pakuta***		

papawitil(pa)	*see* ***kal̲pil-kal̲pil(pa)***		
partjat̲a-partjat̲a [***partjat̲a*** native cat, quoll]	—	*Monachather paradoxus*	grass
pilyal̲i puun(pa) urpa wituka	tarvine and species	*Boerhavia coccinea*	prostrate herb
	• the carrot-like root is baked as an emergency drought food • this is a sticky plant and is used as a tangle-foot to trap small birds		
pulyan̲t̲u	*see* ***mingkul̲(pa)***		
punti	cassia (senna)	*Senna artemisioides* ssp. *sturtii* *Senna artemisioides* ssp. *filifolia*	shrub
	• edible beetle larvae grubs, known as ***maku katat̲i***, are found in the roots		
pur̲ar-pur̲ar(pa)	silver bush	*Ptilotus obovatus*	subshrub
put̲a-put̲a	—	*Cyperus cunninghamii* *Cyperus vaginatus* *Cyperus victoriensis* *Lepidosperma canescens*	sedge
putja nyii-nyii [***putja*** grass (general term), ***nyii-nyii*** zebra finch]	cotton panic grass	*Digitaria brownii*	low grass
puun̲(pa)	*see* ***pilyal̲i***		
puya wintjulany(pa)	bush bean	*Rhyncharrhena linearis*	climbing vine
	• the long, bean-like fruit is cooked and eaten; if food is scarce, the leaves were gathered, ground, baked and eaten		
puyukara kurku	mulga (long, narrow, flat, leaf form B)	*Acacia aneura*	tree
taa-taa	woollybutt wanderrie	*Eriachne helmsii*	tall grass

tawal-tawal(pa)	see **kulypur(pa)**		
tjaamul̲ur̲u	—	*Prostanthera wilkeana*	shrub
tjamalya kurku	mulga (long, narrow, flat leaf form C)	*Acacia aneura*	tree
tjanmat̲a	bush onion	*Cyperus bulbosus*	sedge
	• the bulbs are baked in hot sand and ashes, and eaten, once the outer papery covering is removed		
tjanpi	porcupine grass	*Triodia irritans*	spinifex
tjila walkal(pa)	emu poison bush	*Duboisia hopwoodii*	tall shrub/ small tree
	• the poisonous leaves are dried and crushed, and placed in water holes to stupefy emus and other game; once drugged in this way the animals are easily caught		
	• **Note well**: an extremely poisonous plant — Aboriginal people caution that the plant should not be touched, nor handled near children		
tjilka [*tjilka* thorn, prickle]	—	*Solanum ferocissimum*	subshrub
	cattle bush	*Trichodesma zeylanicum*	woody herb
tjilkal̲a	see **iriya**		
tjilka-tjilka	—	*Sclerolaena diacantha*	prickly subshrub
		Solanum petrophilum	
tjiltjarpi	—	*Hibbertia glaberrima*	spreading shrub
tjilu	see **ilpat̲ilpat̲a**		
tjininy(pa)	hop bush	*Dodonaea viscosa* ssp. *angustissima*	tall shrub
	hop bush	*Dodonaea viscosa* ssp. *mucronata*	shrub
	• a shallow pit is dug, dried leaves are lit in it, and *tjininy(pa)* leaves are put on top to produce smoke; the patient sits over the smoke, lying on one side then the other for relief of the pain		

tjir̲in-tjir̲in(pa)	—	*Abutilon leucopetalum*	shrub
	• children make small play spears with it		
	—	*Abutilon otocarpum*	small velvety shrub
tjul̲pun-tjul̲pun(pa)	wild hops	*Acetosa vesicaria*	leafy subshrub
[**tjul̲pu** bird]	—	*Brunonia australis*	herb
	—	*Chrysocephalum apiculatum*	herb
	paper daisies	*Helipterum* species	herb
	poached egg daisy	*Polycalymma stuartii*	herb
	pussytails	*Ptilotus gaudichaudii* var. *gaudichaudii*	herb
tjuntawar̲a wiriny-wiriny(pa)	bush tomato	*Solanum cleistogamum*	subshrub
	• the ripe, pale yellowish fruits are edible and sweet		
tjuntiwari wan̲ngat̲i	rock isotome	*Isotoma petraea*	herb
	• this is quite a poisonous herb: the sap coming in contact with the eyes can cause temporary blindness		
tjutir̲angu warinkur̲a	—	*Thysanotus exiliflorus*	tuberous lily
tjutiruru	—	*Hibiscus sturtii* var. *grandiflorus*	subshrub
tjut̲u	—	*Eremophila willsii*	low shrub
tulypur(pa)	emu bush	*Eremophila longifolia*	tall shrub
unmut̲a	native cress	*Lepidium muelleri-ferdinandi*	herb
	• steam-baked as an edible green		
	—	*Lepidium phlebopetalum*	herb
un̲talya	*see* ***kur̲umar̲u***		
urpa	*see* ***pilyal̲i***		

utjany(pa)	ironwood • exudes an edible gum called ***tjau***	*Acacia estrophiolata*	tree
walkal(pa)	*see* ***tjila***		
walytjapiri	—	*Tribulus astrocarpus*	flat herb
wanakatakata	*see* ***awalyuru***		
wanngati	*see* ***tjuntiwari***		
warinkura	*see* ***tjutirangu***		
watara	*see* ***munyun(pa)***		
wilypin(pa)	—	*Templetonia egena*	leafless shrub
wilypin-wilypin(pa)	lignum *see also* ***inturalkalpai***	*Muehlenbeckia florulenta*	leafless shrub
wintjulany(pa)	*see* puya		
wintu-wintu	—	*Sida* species synonym *Sida virgata*	spindly shrub
wiriny-wiriny(pa)	*see* ***tjuntawara***		
wituka	*see* ***pilyali***		
yaran(pa)	sandhill grevillea rattlepod grevillea • dried leaves are used as a source of ash for mixing with native tobacco ***ukiri***, ***mingkul(pa)***	*Grevillea stenobotrya*	tall shrub
yuratja	parakeelya	*Calandrinia eremaea*	succulent

List of plants in alphabetical order of botanical name

Genus/species name Family name	Yankunytjatjara name
Abutilon leucopetalum Malvaceae	tjir̲in-tjir̲in(pa)
Abutilon otocarpum Malvaceae	tjir̲in-tjir̲in(pa)
Acacia aneura Mimosaceae	kurku (generic)
long, narrow, flat, leaf form A	kal̲pilya, kurku
long, narrow, flat, leaf form B	puyukara, kurku
long, narrow, flat, leaf form C	tjamalya, kurku
common variable leaf form	wintalyka, kurku
Acacia calcicola Mimosaceae	ikat̲uka
Acacia estrophiolata Mimosaceae	utjany(pa)
Acacia kempeana Mimosaceae	ilykuwara
Acacia ligulata Mimosaceae	wat̲arka
Acacia melleodora Mimosaceae	mintju
Acacia minyura Mimosaceae	minyur̲a, kurku
Acacia murrayana Mimosaceae	tjuntal̲a
Acacia olgana Mimosaceae	kal̲iwar̲a
Acacia ramulosa Mimosaceae	pakut̲a, palpa
Acacia tetragonophylla Mimosaceae	kur̲ara
Acacia victoriae var. *arida* Mimosaceae	ngat̲un̲(pa)
Acetosa vesicaria Polygonaceae	tjul̲pun-tjul̲pun(pa)
Allocasuarina decaisneana Casuarinaceae	kurkar̲a
Aluta maisonneuvei Myrtaceae	pukar̲a, waput̲i
Amphipogon caricinus Poaceae	kutan̲u

Amyema maidenii Loranthaceae	nyinkini nyiruny(pa)
Aristida contorta Poaceae	ipiri, kara
Atriplex nummularia var. *nummularia* Chenopodiaceae	iriya
Atriplex vesicaria Chenopodiaceae	iriya
Boerhavia coccinea Nyctaginaceae	pilyali, puun(pa), urpa, wituka
Brachychiton gregorii Sterculiaceae	ngalta
Brunonia australis Asteraceae	tjulpun-tjulpun(pa)
Calandrinia balonensis Portulacaceae	nyurngi, parkily(pa)
Calandrinia eremaea Portulacaceae	yuratja
Callitris glaucophylla Cupressaceae	kuli, kulilypuru
Calostemma purpureum Amaryllidaceae	apita
Canthium lineare Rubiaceae	awalyuru, pakali-pakali, wanakatakata
Cassia pleurocarpa Caesalpiniaceae	karpil-karpil(pa)
Choiromyces aboriginus Terfeziaceae	witita
Chrysocephalum apiculatum Asteraceae	tjulpun-tjulpun(pa)
Codonocarpus cotinifolius Gyrostemonaceae	kaluti, kanturangu
Convolvulus erubescens Convolvulaceae	anultja
Corymbia opaca Myrtaceae	itara
Crotalaria eremaea ssp. *strehlowii* Caesalpiniaceae	kalpipila
Cymbopogon ambiguus Poaceae	ilintji
Cyperus bulbosus Cyperaceae	tjanmata
Cyperus cunninghamii Cyperaceae	puta-puta
Cyperus vaginatus Cyperaceae	puta-puta

Cyperus victoriensis Cyperaceae	puṯa-puṯa
Dicrastylis costelloi Chloanthaceae	muṟuntu-muṟuntu
Digitaria brownii Poaceae	putja nyii-nyii
Dodonaea viscosa ssp. *angustissima* Sapindaceae	tjininy(pa)
Dodonaea viscosa ssp. *mucronata* Sapindaceae	tjininy(pa)
Duboisia hopwoodii Myoporaceae	tjila, walkal(pa)
Dysphania kalpari Chenopodiaceae	kalpaṟi
Einadia nutans Chenopodiaceae	kaḻpil-kaḻpil(pa)
Einadia nutans ssp. *eremaea* Chenopodiaceae	iwaṯiwata, malkakutjal(pa), mukul-mukul(pa)
Enteropogon acicularis Poaceae	ilintji
Eragrostis eriopoda Poaceae	wanguṉu
Eremophila alternifolia Myoporaceae	irmangka-irmangka
Eremophila duttonii Myoporaceae	munyuṉ(pa)
Eremophila freelingii Myoporaceae	aratja
Eremophila latrobei Myoporaceae	mintjingka, ngaṟankuṟa
Eremophila longifolia Myoperaceae	tulypur(pa)
Eremophila paisleyi Myoporaceae	mani-mani
Eremophila rotundifolia Myoporaceae	aratja
Eremophila sturtii Myoporaceae	munyuṉ(pa)
Eremophila willsii Myoporaceae	tjuṯu
Eriachne helmsii Poaceae	taa-taa
Eucalyptus camaldulensis var. *obtusa* Myrtaceae	apaṟa, iṯaṟa
Eucalyptus eucentrica Myrtaceae	ngapaṟi, tjintjulu

Eucalyptus intertexta Myrtaceae	altar(pa)
Eucalyptus sparsa Myrtaceae	altar(pa)
Eulalia aurea Poaceae	ilintji
Euphorbia drummondii Euphorbiaceae	mangka-mangka
Ficus brachypoda Moraceae	ili
Glycine canescens Papilionaceae	kalpil-kalpil(pa)
Gossypium sturtianum Malvaceae	kalpir-kalpir(pa)
Grevillea juncifolia Proteaceae	ultukun(pa)
Grevillea nematophylla Proteaceae	ilpara
Grevillea stenobotrya Proteaceae	yaran(pa)
Gyrostemon ramulosus Gyrostemonaceae	kurumaru, untalya
Hakea divaricata Proteaceae	witjinti
Hakea lorea ssp. *lorea* Proteaceae	witjinti
Helipterum species Asteraceae	tjulpun-tjulpun(pa)
Hibbertia glaberrima Dilleniaceae	tjiltjarpi
Hibiscus sturtii var. *grandiflorus* Malvaceae	tjutiruru
Isotoma petraea Campanulaceae	tjuntiwari, wanngati
Lepidium muelleri-ferdinandi Brassicaceae	unmuta
Lepidium phlebopetalum Brassicaceae	unmuta
Lepidosperma canescens Cyperaceae	puta-puta
Lysiana exocarpi Loranthaceae	ngantja
Lysiana murrayi Loranthaceae	parka-parka
Melaleuca glomerata Myrtaceae	ilpili

Monachather paradoxus Poaceae	partjata-partjata
Muehlenbeckia florulenta Polygonaceae	wilypin-wilypin(pa)
Mukia maderaspatana Cucurbitaceae	kalpil-kalpil(pa), papawitil(pa)
Nicotiana excelsior Solanaceae	mingkul(pa), pulyantu, ukiri
Olearia stuartii Asteraceae	intiyanu
Pandorea doratoxylon Bignoniaceae	urtjan(pa)
Panicum decompositum Poaceae	kaltu-kaltu, kutja
Pittosporum angustifolium Pittosporaceae	kumpaly(pa)
Podaxis pistillaris Podaxaceae	ilpatilpata, mamawara, tjilu
Polycalymma stuartii Asteraceae	tjulpun-tjulpun(pa)
Portulaca oleracea Portulacaceae	maru-maru, wakati
Prostanthera striatiflora Lamiaceae	karingana
Prostanthera wilkeana Lamiaceae	karingana, tjaamuluru
Pterocaulon sphacelatum Asteraceae	intiyanu
Ptilotus gaudichaudii var. *gaudichaudii* Amaranthaceae	tjulpun-tjulpun(pa)
Ptilotus obovatus Amaranthaceae	purar-purar(pa)
Ptilotus obovatus var. *obovatus* Amaranthaceae	iriya
Ptilotus species Amaranthaceae	liru-liru
Rhagodia eremaea Chenopodiaceae	iwatiwata, malkakutjal(pa), mukul-mukul(pa)
Rhyncharrhena linearis Asclepiadaceae	puya, wintjulany(pa)
Salsola tragus Chenopodiaceae	iriya, tjilkala
Santalum acuminatum Santalaceae	mangata
Santalum lanceolatum Santalaceae	kupata

Sarcostemma viminale ssp. *australe* Asclepiadaceae	ipi-ipi
Sclerolaena diacantha Chenopodiaceae	tjilka-tjilka
Senecio magnificus Asteraceae	liru-liru
Senna artemisioides ssp. *filifolia* Caesalphiniacea	punti
Senna artemisioides ssp. *helmsii* Caesalpiniaceae	aripita
Senna artemisioides ssp. *sturtii* Caesalpiniaceae	punti
Sida filiformis Malvaceae	ngau-ngau
Sida species Malvaceae	wintu-wintu
Solanum centrale Solanaceae	kampur̲ara, kampur̲ar(pa)
Solanum cleistogamum Solanaceae	tjuntawar̲a, wiriny-wiriny(pa)
Solanum coactiliferum Solanaceae	it̲uny(pa), kumpul(pa)
Solanum ellipticum Solanaceae	kulypur(pa), tawal-tawal(pa)
Solanum ferocissimum Solanaceae	tjilka
Solanum petrophilum Solanaceae	tjilka-tjilka
Spartothamnella teucriiflora Chloanthaceae	intur̲alkalpai, wilypin-wilypin(pa)
Stemodia florulenta Scrophulariaceae	intiyanu
Stemodia viscosa Scrophulariaceae	intiyanu
Templetonia egena Papillionaceae	wilypin(pa)
Themeda australis Poaceae	ilintji
Thysanotus exiliflorus Liliaceae	tjutir̲angu, warinkur̲a
Tribulus astrocarpus Zygophyllaceae	walytjapiri
Trichodesma zeylanicum Boraginaceae	tjilka
Triodia irritans Poaceae	tjanpi

References and Resources

Angatja Video and Riverbed Productions. 1984. *Mayi Wiṟu: It's great food.* (Video, with English subtitles, of traditional winter foods. Featuring Nganintja.)

Areyonga Literature Production Centre. 1983– . *Tjakulpa Mulapa.* Areyonga, Northern Territory. (A community magazine; includes a series of articles on traditional Pitjantjatjara bush foods, in the Pitjantjatjara language, with illustrations and black-and-white photographs.)

The Australian Systematic Botany Society. 1981. *Flora of Central Australia* (John Jessop, Editor-in-chief). Sydney: Reed.

Basedow, H. 1904. 'Anthropological notes made on the South Australian Government North-West Prospecting Expedition, 1903'. Transactions Royal Society South Australia. 28: 12–51. (Names recorded in the vocabulary of the 'Karkurrerra tribe' inhabiting 'the greater portion of the Musgrave Ranges, more particularly the southern limits'.)

Black, J.M. 1915. 'Botany', Transactions Royal Society South Australia. 39:823–42 (Names and information from the 1914 geological expedition into the Musgrave and Everard Ranges.)

Black, J.M. 1915. 'Language of the Everard Range Tribe', Transactions Royal Society South Australia: 39: 732–5 (Names and information as in the preceeding.)

Brokensha, P. 1978. *The Pitjantjatjara and their Crafts.* Sydney: Aboriginal Arts Board, Australia Council.

Bryce, S. 1983. 'The role of bush-tucker in nutrition education'. In K. O'Dea (ed), 20–3.

Bryce, S. 1985. *A Health Worker's Nutrition Handbook for Central Australia.* Alice Springs: Institute for Aboriginal Development. (English and Pitjantjatjara editions.)

Bryce, S. 1986. *Women's Gathering and Hunting in the Pitjantjatjara Homelands.* A Resource Book with Slides. Alice Springs: Institute for Aboriginal Development. (Twenty-nine items of bushfood illustrated and described, an account of women's traditional economic role, and those aspects maintained today.)

Cane, S. 1985. 'Bush Tucker: Intensified use of traditional resources on Aboriginal outstations'. In B. Foran and B. Walker (eds), *Science and Technology for Aboriginal Development.* Melbourne: CSIRO.

Cleland, J.B. And Johnston, T.H. 1937–1938. 'Notes on native names and uses of plants in the Musgrave Ranges Region'. *Oceania* 8(2): 208–15 and 8(3): 328–42. (Specific Yankunytjatjara names and information recorded.)

Douglas, W.H. 1977. *Illustrated Topical Dictionary of the Western Desert Language.* Australian Institute of Aboriginal Studies, Canberra. (Available in the Warburton dialect, and also as a blank version where specific local dialect names may be entered.)

Ernabella Video Productions. 1985. *Bush Medicine.* (Video, with English subtitles, featuring Nora Rupert, Nura Ward, Anmanari Alice.)

Gallagher, P. And Poulson, M. 1983. *Mai Tjuratja*. Areronga Literature Production Centre. Ayeronga, Northern Terriroty. (An illustrated booklet on honey plants, and honey from bees and ants, in the Pitjantjatjara language, with illustrations.)

Goddard, C. 1980. *A Learner's Guide to Yankunytjatjara*. Alice Springs: Institute for Aboriginal Development.

Goddard, C. 1984. *Yankunytjatjara Grammar*. Alice Springs: Institute for Aboriginal Development.

Goddard, C. 1987. *A Basic Pitjantjatjara/Yankunytjatjara to English Dictionary*. Alice Springs: Institute for Aboriginal Development.

Griffin, G.F. 1985. 'Bush-tucker: Grow your own'. *Australian Plants* 13(103), June 1985, 127–34.

Hamilton, A. 1980. 'Dual Social Systems: Technology, Labour and Women's Secret Rites in the Eastern Western Desert of Australia'. *Oceania*. LI(1), 4–19.

Helms, R. 1896. 'Anthropology. Report of Elder Scientific Expedition 1891'. Transactions Royal Society South Australia 16(3): 237–332. (Names recorded in 'Vocabulary obtained from natives of the Everard Range tribe', p318 — some names, not included in the above checklist, recorded without botanical names.)

Hetzel, B.S. And Frith, H.J. (eds). 1982. *The Nutrition of Aborigines in relation to the Ecosystem of Central Australia*. Melbourne: CSIRO. (Papers presented at a symposium, CSIRO 23–26 Oct 1976.)

Kalotas, A.C. 1983. 'A list of plant species traditionally and currently utilised by the Pitjantjatjara people of Pipalyatjara and the surrounding region'. In H.C. Coombs, M.M. Brandl, W.E. Snowden (eds), *A Certain Heritage: Programs for and by Aboriginal families in Australia*. Canberra: Centre for Resource and Environmental Studies.

Kalotas, A.C. 1983. 'Learning about plant foods in Pitjantjatjara areas'. In K.O'Dea (ed), 28–31.

King, P. (ed). Plant Identikit. Alice Springs: Conservation Commission of the N.T.

Latz, P.K. 1995. *Bushfires and Bushtucker: Aboriginal Plant Use in Central Australia*. Alice Springs: IAD Press.

Latz, P.K. and Griffin, G.F. 1982. 'Changes in Aboriginal land management in relation to fire, and to food-plants in Central Australia'. In B.S. Hetzel and H.J. Frith (eds), 77–85.

Lester, Y. 1982. Pages from an Aboriginal Book. Alice Springs: Institute for Aboriginal Development.

O'Dea, K. (ed). 1983. Proceedings, Aboriginal Bushfoods Workshop, Griffith University, Brisbane, November 19–20. Brisbane: Griffith University.

Peterson, N. 1978. 'The traditional pattern of subsistence to 1975'. In B.S. Hetzel and H.J. Frith (eds), 25–35.

Reid, J. (ed). 1982. Body, Land and Spriit: Health and Healing in Aboriginal Society. St Lucia, Qld: University of Queensland Press.

Tindale. N.B. 1941. 'A list of plants collected in the Musgrave and Mann Ranges, South Australia, 1933'. *The South Australian Naturalist* 21(1): 8–12. (Specific Yankunytjatjara names and information recorded.)

Silberbauer 1971, 'Ecology of the Ernabella Aboriginal community'. Anthropological Forum 3(1): 21–36 (Specific Yankunytjatjara names and information recorded.)

Whibley, D.J.E. 1980. *Acacias of South Australia*. Adelaide: Government Printer.

Winfield, C. 1982. Bush Tucker. A guide to and resources on traditional Aboriginal foods in the North West of South Australia and Central Australia. Adelaide: Wattle Park Teachers Centre. (Specific Yankunytjatjara names and information recorded. Illustrated.)

Index of common, botanical and Yankunytjatjara names

Note: bold page numbers indicate main entry

buckbush (*Salsola tragus*) 97, 109
bush bean vine (*Rhyncharrhena linearis*) 5, 9, 101, 109
bush currant (*Canthium lineare*) 5, 8, 9, 96, 106
bush onion (*Cyperus bulbosus*) 4, 11, 102, 106
bush plum (*Santalum lanceolatum*) **50–2**, 109
bush tomato (various *Solanum* species) 5, 8, 9, **61–3**, 89, 93, 98, 103, 110
Calandrinia balonensis (see also parakeelya) **67–8**, 106
Calandrinia eremaea (see also parakeelya) 104, 106
Callitris glaucophylla (see also native pine) 98, 106
Calostemma purpureum (garland lily) 96, 106
Canthium lineare (see also bush currant) 96, 106
cassia (senna) 4, 5, 89, 96, 98, 101, 106
Cassia pleurocarpa 98, 106
cattle bush (*Trichodesma zeylanicum*) 102, 110
caustic vine (*Sarcostemma viminale* var. *australe*) 3, 7, 13, 15, **72–3**, 110
Choiromyces aboriginus (see also native truffle) **87–8**, 106
Chrysocephalum apiculatum 103, 106
Codonocarpus cotinifolius (see also desert poplar) 98, 106
colony wattle (*Acacia murrayana*) 3, 5, 10, **52–4**, 90, 105
Convolvulus erubescens (bindweed) 96, 106
corkwood (*Hakea lorea* ssp. *lorea*, *H. divaricata*; see also ***witjinti***) 2, 4, 5, 7, 12, 13, **38–9**, 89, 90, 91, 108
Corymbia opaca (see also bloodwood) 97, 106
cotton panic grass (*Digitaria brownii*) 101, 107
Crotalaria eremaea ssp. *strehlowii* (see also ***kalpipila***) **63–4**, 106
curly wire grass (*Aristida contorta*) 97, 106
Cymbopogon ambiguus (scent grass) 96, 106
Cyperus bulbosus (see also bush onion) 102, 106
Cyperus cunninghamii 101, 106
Cyperus vaginatus 101, 106
Cyperus victoriensis 101, 107
daisies (*Helipterum* species) 103, 108
dead finish (*Acacia tetragonophylla*) 4, 10, 99, 105
desert kurrajong (*Brachychiton gregorii*) 6, 100, 106
desert mulga (*Acacia minyura*) 14, **31–3**, 105
desert oak (*Allocasuarina decaisneana*) 6, 99, 105
desert poplar (*Codonocarpus cotinifolius*) 5, 98, 108
desert raisin (*Solanum centrale*) 5, 8, 9, 93, 98, 110
desert thryptomene (*Aluta maisonneuvei*) 3, 5, 12, **56–8**, 105
Dicrastylis costelloi 100, 107
Digitaria brownii (cotton panic grass) 101, 107
Dodonaea viscosa ssp. *angustissima* (see also hop bush) 102, 107
Dodonaea viscosa var. *mucronata* (see also hop bush) 102, 107
Duboisia hopwoodii (see also emu poison bush) 102, 107
Dysphania kalpari (see also rats' tails) 98, 107
Einadia nutans 98, 107
Einadia nutans ssp. *eremaea* 97, 107
emu bush (*Eremophila longifolia*) 4, 103, 107
emu poison bush (*Duboisia hopwoodii*) 5, 14, 102, 107
Enteropogon acicularis (umbrella grass) 96, 107
Eragrostis eriopoda (see also naked woollybutt) **85–7**, 107
Eremophila alternifolia (see also native fuchsia) **46–8**, 107

Eremophila duttonii (red poverty bush) 100, 107

Eremophila freelingii (see also native fuchsia) **40–1**, 107

Eremophila latrobei (see also fuchsia bush, native fuchsia) 100, 107

Eremophila longifolia (see also emu bush) 103, 107

Eremophila paisleyi (see also native fuchsia) 99, 107

Eremophila rotundifolia (see also native fuchsia) 96, 107

Eremophila sturtii (turpentine bush) 100, 107

Eremophila willsii 103, 107

Eriachne helmsii (woollybutt wanderrie) 101, 107

Eucalyptus camaldulensis var. *obtusa* (see also river red gum) **25–6**, 107

Eucalyptus eucentrica (water mallee, red mallee) 100, 107

Eucalyptus intertexta (gum-barked coolibah) 96, 108

Eucalyptus sparsa (gum-barked coolibah) 96, 108

Eulalia aurea (silky brown top) 96, 108

Euphorbia drummondii (matspurge, milk weed) 99, 108

Ficus brachypoda (see also wild fig) 96, 108

fuchsia bush (see also *Eremophila latrobei*) 4

garland lily (*Calostemma purpureum*) 96, 106

Glycine canescens 98, 108

Gossypium sturtianum (Sturt's desert rose) 98, 108

Grevillea juncifolia (see also honeysuckle grevillea) **54–5**, 108

Grevillea nematophylla (see also silver-leaved water bush) **42–4**, 108

Grevillea stenobotrya (see also sandhill grevillea; rattlepod grevillea) 104, 108

greybeard grass (*Amphipogon caricinus*) **83–4**, 105

Gyrostemon ramulosus (see also wheel fruit) 5, 99, 108

gum-barked coolibah (*Eucalyptus intertexta*) 96, 108

Hakea divaricata (see also corkwood) **36–7**, 108

Hakea lorea ssp. *lorea* (see also corkwood) **36–7**, 108

Helipterum species 103, 108

Hibbertia glaberrima 102, 108

Hibiscus sturtii var. *grandiflorus* 103, 108

honeysuckle grevillea (*Grevillea juncifolia*) 3, 5, 12, **54–5**, 108

hop bush (*Dodonea viscosa* ssp. *angustissima*, *D. viscosa* ssp. *mucronata*) 3, 4, 5, 13, 102, 107

horse mulga (*Acacia ramulosa*) 100, 105

ikaṯuka (*Acacia calcicola*) 96, 105

iḻi (wild fig) 2, 3, 6, 8, 9, 96, 108

ilintji (tall grasses) 4, 7, 96, 106, 107, 108, 110

iḻpaṟa (silver-leaved water bush) 4, 15, **42–4**, 108

ilpaṯilpaṯa (stalked puffball) 15, 96, 109

ilpili (inland ti-tree) 4, 97, 108

ilykuwara (witchetty bush) 2, 4, 5, **44–6**, 105

inland paperbark (*Melaleuca glomerata*; see also inland ti-tree) 97, 108

inland pigweed (*Portulaca oleracea*) 5, 10, **69–71**, 89, 109

inland ti-tree (*Melaleuca glomerata*) 4, 97, 108

intiyanu (blue rod; apple bush; *Olearia stuartii*; *Stemodia florulenta*) 4, 97, 109, 110

intuṟalkalpai (*Spartothamnella teucriiflora*) 97, 110

ipi-ipi (caustic vine) 3, 7, 13, 15, **72–3**, 110

ipiṟi (curly wire grass) 97, 106

iriya (saltbush; silver bush; rolypoly, buckbush) 97, 106, 109

irmangka-irmangka (*Eremophila alternifolia*) 3, 13, **46–8**, 107

ironwood (*Acacia estrophiolata*; see also ***utjany(pa)***) 2, 4, 5, 12, 104, 105

Isotoma petraea (see also rock isotome) 103, 108

itara (bloodwood) 2, 4, 5, 7, 15, 97, 106, 107

itara (river red gum) (see ***apara***)

ituny(pa) (western nightshade) 5, 8, 9, **61–3**, 89, 110

iwatiwata (*Einadia nutans* ssp. *eremaea*; *Rhagodia eremaea*) 97, 107, 109

kaliwara (*Acacia olgana*) 3, 10, 97, 105

kalpari (rats' tails) 5, 10, 98, 107

kalpil-kalpil(pa) (*Einadia nutans*; *Glycine canescens*; *Mukia maderaspatana*) 98, 107, 108, 109

kalpilya (mulga) 98, 105

kalpipila (*Crotalaria eremaea* ssp. *strehlowii*) 13, **63–4**, 106

kalpir-kalpir(pa) (Sturt's desert rose) 98, 108

kaltu-kaltu (native millet) 5, 7, 10, **81–2**, 109

kaluti (desert poplar) 5, 98, 106

kampurar(pa) (see ***kampurara***)

kampurara (desert raisin) 5, 8, 9, 98, 110

kangaroo grass (*Themeda australis*) 96, 110

kanturangu (see ***kaluti***)

kara (see also ***ipiri***) 106

karingana (mint bush) 3, 13, **49**, 109

karpil-karpil(pa) (*Cassia pleurocarpa*) 98, 106

kuli (native pine; see also ***kulilypuru***) 98, 106

kulilypuru (native pine) 2, 4, 98, 106

kulypur(pa) (wild gooseberry) 9, 99, 110

kumpaly(pa) (native willow) 90, 99, 109

kumpul(pa) (see ***ituny(pa)***)

kupata (wild plum, bush plum) 4, 5, 8, 9, **50–2**, 109

kurara (dead finish) 4, 10, 99, 105

kurkara (desert oak) 6, 99, 105

kurku (mulga) 2, 4, 7, 8, 10, 11, 12, 14, 15, **31–7**, 76, 89, 90, 91, 98, 100, 101, 102, 105

kurumaru (wheel fruit) 5, 99, 108

kutanu (greybeard grass) **83–4**, 105

kutja (see ***kaltu-kaltu***)

Lepidium muelleri-ferdinandi (see also native cress) 103, 108

Lepidium phlebopetalum (see also native cress) 103, 108

Lepidosperma canescens 101, 108

lignum (*Muehlenbeckia florulenta*) 104, 109

liru-liru (pussytails; tall yellow top) 99, 109, 110

Lysiana exocarpi (see also mistletoe) **74–5**, 108

Lysiana murrayi (see also mistletoe) **74–5**, 108

malkakutjal(pa) (*Einadia nutans* ssp. *eremaea*; *Rhagodia eremaea*) (see ***iwatiwata***)

mallee 5, 7, 96, 100, 107–8

mamawara (see ilpatilpata)

mangata (quandong) 8, 9, 13, 15, **27–30**, 90, 109

mangka-mangka (matspurge, milk weed) 99, 108

mani-mani (*Eremophila paisleyi*) 99, 107

maru-maru (see ***wakati***)

matspurge (*Euphorbia drummondii*) 99, 108

Melaleuca glomerata (see also inland ti-tree; inland paperbark) 97, 108

milk weed (*Euphorbia drummondii*) 99, 108

mingkul(pa) (see also wild tobacco) 4, 14, 25, **65–7**, 91, 109

mint bush (*Prostanthera striatiflora*) 3, 13, **49**, 109

mintjingka (*Eremophila latrobei*) 4, 12, 100, 107

mintju (*Acacia melliodora*) 100, 105

minyura (desert mulga) 14, **31–3**, 105

mistletoe (*Amyema maidenii; Lysiana exocarpi; L. murrayi*) 8, 9, **74–5**, 100, 106, 108

Monachather paradoxus 101, 109

Muehlenbeckia florulenta (lignum) 104, 109

Mukia maderaspatana 98, 109

mukul-mukul(pa) (see also ***iwatiwati***) 7

mulga (various *Acacia* species) 2, 4, 5, 7, 8, 10, 11, 12, 14, 15, 76, 89, 90, 91, 98, 100, 101, 102, 105

munyun(pa) (red poverty bush; turpentine bush) 13, 100, 107

muruntu-muruntu (*Dicrastylis costelloi*) 100, 107

naked woollybutt (*Eragrostis eriopoda*) 5, 10, **85–7**, 107

native cress (*Lepidium muelleri-ferdinandi*) 11, 103, 108

native fuchsia (various *Eremophila* species) 2, 3, 5, 12, 13, **40–1**, **46–8**, 96, 99, 100, 107

native millet (*Panicum decompositum*) 5, 7, 10, **81–2**, 109

native pine (*Callitris glaucaphylla*) 2, 4, 98, 106

native tobacco (see also wild tobacco) 6, 91

native truffle (*Choiromyces aboriginus*) 5, 11, **87–8**, 106

native willow (*Pittosporum angustifolium*) 90, 99, 109

ngalta (desert kurrajong) 6, 100, 106

ngantja (see also ***parka-parka***) 108

ngapari (sweet secretion) 12, 25, 26, 90

ngapari (water mallee, red mallee) 100, 107

ngarankura (see ***mintjingka***)

ngatun(pa) (prickly wattle, acacia bush) 4, 10, 100, 105

ngau-ngau (*Sida filiformis*) 100, 110

Nicotiana excelsior (see also wild tobacco) **65–7**, 109

nyinkini (*Amyema maidenii*) 100, 106

nyiruny(pa) (see ***nyinkini***)

nyurngi (see ***parkily(pa)***)

old man saltbush (*Atriplex nummularia* var. *nummularia*) 97, 106

Olearia stuartii 97, 109

pakali-pakali (see also ***awalyuru***) 5, 7, 8, 9

pakuta (horse mulga) 100, 105

palpa (see ***pakuta***)

Pandorea doratoxylon (see also spear bush) **76–80**, 109

Panicum decompositum (see also native millet) **81–2**, 109

papawitil(pa) (see ***kalpil-kalpil(pa)***)

paper daisies (*Helipterum* species) 103, 108

parakeelya (*Calandrinia balonensis*, *C. eremaea*) 5, 11, **67–8**, 104, 106

parka-parka (*Lysiana exocarpi*, *L. murrayi*) 9, **74–5**, 108

parkily(pa) (parakeelya) 5, 11, **67–8**, 104, 106

partjata-partjata (*Monochather paradoxus*) 101, 109

pigweed (see inland pigweed)

pilyali (tarvine) 11, 101, 106

Pittosporum angustifolium (native willow) 99, 109

pituri (see wild tobacco)

poached egg daisy (*Polycalymma stuartii*) 103, 109

Podaxis pistillaris (see also stalked puffball) 96, 109

Polycalymma stuartii (poached egg daisy) 103, 109

porcupine grass (*Triodia irritans*) 102, 110

Portulaca oleracea (see also inland pigweed) **69–71**, 109

prickly wattle (*Acacia victoriae* var. *arida*) 4, 10, 100, 105

Prostanthera striatiflora (see also mint bush) **49**, 109

Prostanthera wilkeana 102, 109

Pterocaulon sphacelatum (apple bush) 97, 109

Ptilotus gaudichaudii var. *gaudichaudii* (pussytails) 103, 109

Ptilotus obovatus (silver bush) 101, 109

Ptilotus obovatus var. *obovatus* (silver bush) 97, 109

Ptilotus species (pussytails) 99, 103, 109

pukar̠a (see ***waput̠i***)

pulyan̠t̠u (see also ***mingkul̠(pa)***) 4

punti (*Senna* species) 4, 5, 101, 110

pur̠ar-pur̠ar(pa) (silver bush) 101, 109

pussytails (*Ptilotus* species) 7, 99, 103, 109

put̠a-put̠a (various sedges) 4, 6, 7, 101, 106, 107, 108

putja nyii-nyii (cotton panic grass) 101, 107

puun̠(pa) (see ***pilyal̠i***)

puya (bush bean vine) 101, 109

puyukara (mulga) 101, 105

quandong (*Santalum acuminatum*) 8, 9, 13, 15, **27–30**, 90, 109

rats' tails (*Dysphania kalpari*) 5, 10, 98, 107

rattlepod grevillea (see sandhill grevillea)

red mallee (*Eucalyptus eucentrica*) 100, 107

red poverty bush (*Eremophila duttonii*) 100, 107

Rhagodia eremaea 97, 109

Rhyncharrhena linearis (see also bush bean vine) 101, 109

river red gum (*Eucalyptus camaldulensis* var. *obtusa*) 3, 4, 7, 12, 14, 15, **25–6**, 89, 90, 107

rock isotome (*Isotoma petraea*) 4, 103, 108

rolypoly (*Salsola tragus*) 97, 109

Salsola tragus (rolypoly, buckbush) 97, 109

sandhill grevillea (*Grevillea stenobotrya*) 3, 5, 14, 104, 108

Santalum acuminatum (see also quandong) **27–30**, 109

Santalum lanceolatum (bush plum, wild plum) **50–2**, 109

Sarcostemma viminale ssp. *australe* (see also caustic vine) **72–3**, 110

scent grass (*Cymbopogon ambiguus*) 96, 106

Sclerolaena diacantha 102, 110

Senecio magnificus (tall yellow top) 99, 110

senna (cassia) 4, 5, 89, 96, 98, 101, 106

Senna artemisioides ssp. *filifolia* 101, 110

Senna artemisioides ssp. *helmsii* 96, 110

Senna artemisioides ssp. *sturtii* 101, 110

Sida filiformis 100, 110

Sida species 104, 110

silky brown top (*Eulalia aurea*) 96, 108

silver bush (*Ptilotus obovatus*, *P. obovatus* var. *obovatus*) 97, 101, 109

silver-leaved water bush (*Grevillea nematophylla*) 4, 15, **42–4**, 108

Solanum centrale (see also desert raisin) 98, 110

Solanum cleistogamum (see also bush tomato) 103, 110

Solanum coactiliferum (see also western nightshade, bush tomato) **61–3**, 110

Solanum ellipticum (wild gooseberry) 99, 110

Solanum ferocissimum 102, 110

Solanum petrophilum 102, 110

Spartothamnella teucriiflora 97, 110

spear bush (*Panicum decompositum*) 4, 14, **76–80**, 90, 109

spinifex (*Triodia* species) 2, 3, 4, 5, 14, 91, 92, 102, 110

stalked puffball (*Podaxis pistillaris*) 15, 96, 109

Stemodia florulenta 97, 110

Stemodia viscosa (see also blue rod) 97, 110

Sturt's desert rose (*Gossypium sturtianum*) 98, 108

taa-taa (woollybutt wanderrie) 101, 107

tall yellow top (*Senecio magnificus*) 99, 110

tarvine (*Boerhavia coccinea*) 11, 101, 106

tawal-tawal(pa) (see ***kulypur(pa)***)

Templetonia egena 104, 110

Themeda australis (kangaroo grass) 96, 110

Thysanotus exiliflorus 103, 110

tjaamuluru (*Prostanthera wilkeana*) 102, 109

tjamalya (mulga) 102, 105

tjanmata (bush onion) 4, 11, 102, 106

tjanpi (porcupine grass; spinifex) 2, 3, 14, 102, 110

tjila (emu poison bush) 5, 14, 102, 107

tjilka (cattle bush; *Solanum ferocissimum*) 7, 102, 110

tjilka (prickle) 91

tjilkala (see ***iriya***)

tjilka-tjilka (*Sclerolaena diacantha; Solanum petrophilum*) 7, 102, 110

tjiltjarpi (*Hibbertia glaberrima*) 102, 108

tjilu (see ***ilpatilpata***)

tjininy(pa) (hop bush) 3, 4, 5, 13, 102, 107

tjintjulu (see ***ngapari***)

tjirin-tjirin(pa) (*Abutilon leucopetalum; Abutilon otocarpum*) 103, 105

tjulpun-tjulpun(pa) (various wildflower species) 7, 103, 105, 106, 108, 109

tjuntala (colony wattle) 3, 5, 10, **52–4**, 90, 105

tjuntawara (bush tomato) 103, 110

tjuntiwari (rock isotome) 4, 103, 108

tjutirangu (*Thysanotus exiliflorus*) 103, 110

tjutiruru (*Hibiscus sturtii* var. *grandiflorus*) 103, 108

tjutu (*Eremophila willsii*) 5, 103, 107

Tribulus astrocarpus 104, 110

Trichodesma zeylanicum (cattle bush) 102, 110

Triodia irritans (porcupine grass; see also spinifex) 102, 110

tulypur(pa) (emu bush) 4, 103, 107

turpentine bush (*Eremophila sturtii*) 13, 100, 107

ukiri (see ***mingkul(pa)***)

ultukun(pa) (honeysuckle grevillea) 3, 5, 12, **54–5**, 89, 108

umbrella bush (*Acacia ligulata*) 4, 5, **59–60**, 105

umbrella grass (*Enteropogon acicularis*) 96, 107

unmuta (native cress) 11, 103, 108

untalya (wheel fruit) 5, 99, 108

urpa (see ***pilyali***)

urtjan(pa) (spear bush) 4, 14, **76–80**, 90, 109

utjany(pa) (ironwood) 2, 4, 5, 12, 104, 105

wakati (inland pigweed) 5, 10, **69–71**, 89, 109

walkal(pa) (see also ***tjila***) 5

walytjapiri (*Tribulus astrocarpus*) 104, 110

wanakatakata (see ***awalyuru***)

wangunu (naked woollybutt) 5, 10, **85–7**, 107

wanngati (see ***tjuntiwari***)

waputi (desert thryptomene) 3, 5, 12, **56–8**, 105

warinkura (see ***tjutirangu***)

watara (see ***munyun(pa)***)

watarka (*Acacia ligulata*) 4, 5, **59–60**, 105

water mallee (*Eucalyptus eucentrica*) 100, 107

wattle (see acacia)

wayanu (see ***mangata***)

western nightshade (*Solanum coactiliferum*) 5, 8, 9, **61–3**, 110

wheel fruit (*Gyrostemon ramulosus*) 5, 99, 108

wild fig (*Ficus brachypoda*) 2, 3, 6, 8, 9, 96, 108

wild gooseberry (*Solanum ellipticum*) 8, 99, 110

wild hops (*Acetosa vesicaria*) 103, 105

wild plum (*Santalum lanceolatum*) 4, 5, 8, 9, **50–2**, 109

wild tobacco (*Nicotiana excelsior*; see also native tobacco) 4, 14, 25, **65–7**, 109

wilypin(pa) (*Templetonia egena*) 104, 110

wilypin-wilypin(pa) (lignum; *Spartothamnella teucriiflora*) 97, 104, 109, 110

wintalyka (mulga) 10, **33–7**, 105

wintjulany(pa) (bush bean vine; see also ***puya***) 5, 9, 101, 109

wintu-wintu (*Sida* species) 104, 110

wiriny-wiriny(pa) (see ***tjuntawaṟa***)

witchetty bush (*Acacia kempeana*) 2, 4, 5, **44–6**, 105

witiṯa (native truffle) 5, 11, **87–8**, 106

witjinti (corkwood) 2, 4, 5, 7, 12, 13, **38–9**, 89, 90, 91, 108

wituka (see ***pilyaḻi***)

woollybutt wanderrie (*Eriachne helmsii*) 101, 107

yaṟan(pa) (sandhill grevillea, rattlepod grevillea) 3, 5, 14, 104, 108

yuratja (see also ***parkily(pa)***) 104, 106